ROADS TO COMMENSURABILITY

SYNTHESE LIBRARY

STUDIES IN EPISTEMOLOGY,

LOGIC, METHODOLOGY, AND PHILOSOPHY OF SCIENCE

Managing Editor:

JAAKKO HINTIKKA, *Florida State University, Tallahassee*

Editors:

DONALD DAVIDSON, *University of California, Berkeley*
GABRIËL NUCHELMANS, *University of Leyden*
WESLEY C. SALMON, *University of Pittsburgh*

VOLUME 187

DAVID PEARCE

Freie Universität Berlin

ROADS TO
COMMENSURABILITY

D. REIDEL PUBLISHING COMPANY

A MEMBER OF THE KLUWER ACADEMIC PUBLISHERS GROUP

DORDRECHT / BOSTON / LANCASTER / TOKYO

Library of Congress Cataloging–in–Publication Data

CIP

Pearce, David, 1952–
   Roads to commensurability.

   (Synthese library ; v. 187)
   Revision of the author's Habilitationsschrift,
Freie Universität, Berlin, 1986.
   Bibliography: p.
   Includes indexes.
   1. Science—Philosophy. I. Title. II. Title:
Commensurability.
Q175.P349   1987   501   87–3012
ISBN 90–277–2414–8

Published by D. Reidel Publishing Company,
P.O. Box 17, 3300 AA Dordrecht, Holland.

Sold and distributed in the U.S.A. and Canada
by Kluwer Academic Publishers,
101 Philip Drive, Norwell, MA 02061, U.S.A.

In all other countries, sold and distributed
by Kluwer Academic Publishers Group,
P.O. Box 322, 3300 AH Dordrecht, Holland.

Printed in The Netherlands

For Anita

CONTENTS

PREFACE

How many miles to Babylon?
Three-score and ten.
Can I get there by candle-light?
Yes, and back again.
If your heels are nimble and light,
You may get there by candle-light.

Any philosopher who takes more than a fleeting interest in the sciences and their development must at some stage confront the issue of incommensurability in one or other of its many manifestations. For the philosopher of science concerned with problems of conceptual change and the growth of knowledge, matters of incommensurability are of paramount concern. After many years of skating over, skimming through and skirting round this issue in my studies of intertheory relations in science, I decided to take the plunge and make the problem of incommensurability the central and unifying theme of a book. The present volume is the result of that decision.

My interest in problems of comparability and commensurability in science was awakened in the formative years of my philosophical studies by my teacher, Jerzy Giedymin. From him I have learnt not only to enjoy philosophical problems but also to beware of simpleminded solutions to them. The vibrant seminars of Paul Feyerabend held at Sussex University in 1974 left me in no doubt that incommensurability was, and would remain, a major topic of debate and dispute in the philosophical study of human knowledge. More recently, a sustained period of collaborative research with Veikko Rantala on the structure of scientific

ix

theories and their interrelations has provided me with the impetus and the instruments for embarking on the present enterprise. A substantial part of this book is in fact a direct outgrowth of the joint programme in metascience that we have been engaged on since 1980.

My debt to Veikko Rantala is indeed a very special one. He has been a constant source of ideas, advice and encouragement throughout the period of writing this work. He has read and commented on several earlier versions of the manuscript. And, above all, the last four chapters draw freely and frequently on the results of our joint programme. In fact, the volume as a whole has very much the character of a companion to our forthcoming book on the correspondence relation, referred to here as Pearce and Rantala (1988). For his friendship and untiring intellectual support I wish to express to Veikko my warmest thanks.

In 1986, the penultimate version of the text was circulated among several friends and colleagues, and was submitted to the Freie Universität Berlin as my *Habilitationsschrift*. For their constructive criticisms which have led, I hope, to considerable improvements in the final version, I am grateful to Peter Krausser, Lorenz Krüger and Hans Rott.

I also wish to thank Wolfgang Balzer and Wolfgang Stegmüller whose own works figure so prominently in the first two chapters of this book. Despite my sometimes over-spirited criticisms of their arguments, they have been patient and receptive in dealing with my views, tolerant and generous in responding to them. To Wolfgang Balzer I am also grateful for the many friendly discussions we have had, in person and in correspondence.

This project might have faltered at a very early stage, were it not for the valuable encouragement of many persons. Here a special mention should be made of Ilkka Niiniluoto, who at various stages offered numerous suggestions for improving the text; Jaakko Hintikka, who welcomed the project, advised in the course of its development, and supported its publication in *Synthese Library*; and Mrs J. C. Kuipers of Reidel, who has skilfully co-ordinated all the matters involved in bringing a text from rough draft to printed page.

Preliminary work on parts of this book was carried out in the congenial atmosphere of the Netherlands Institute for Advanced Studies (NIAS), in Wassenaar, Holland, where I held a Fellowship during 1982—83. A few passages from Chapter 1 have been taken from my paper (1982b), published in *The British Journal of the Philosophy of Science*; a shorter version of Chapter 2 has recently appeared in *Erkenntnis* (Pearce, 1986); and Chapter 3 overlaps with my (1984), published in the *Zeitschrift für allgemeine Wissenschaftstheorie*. For permission to make use of this material here, I am grateful to the all editors and publishers concerned.

Some indented expressions in the text are numbered for subsequent reference. When such an expression is referred to in a later chapter the number in question is prefixed by the number of the chapter in which it first occurs.

Berlin, March, 1987

INTRODUCTION

The doctrine of incommensurability needs no introduction. It is now twenty-five years since Paul Feyerabend and Thomas Kuhn gave the word 'incommensurable' a new lease of life, using it to describe the situation that occurs when scientific theories and even richer systems of belief are conceptually very distant from one another. Having coined the expression 'incommensurable' to apply to such pairs of apparently conflicting yet not strictly rationally comparable theories, conceptual frameworks and Weltanschauungen, they went on to argue for their celebrated thesis that examples of this phenomenon abound in the history of science and in the history of rational thought generally.

This is a radical claim which caught the mood of the sixties. Targeted on the dominant empiricist and critical rationalist epistemologies of science, its effect was to create a polarisation in the philosophical community. Arguments for and against it were hurled with unusual venom across what came to be the Great Divide. On the one side were philosophers who sought to preserve the sturdy values of rationality and progress and to defend the cognitive elements of science and its methodology. On the other side gathered philosophers with a flair for detecting the metaphysical and the mystical even among the more sober scientific pursuits, and who delighted in the chance to put down the nobler rationalist categories, such as 'logic' and 'experience', and, ironically, clothe their arguments in authoritative and 'scientific' dress at the same time. Whatever the origins and aims of Kuhn's and Feyerabend's doctrine, it is plain that their thesis gave fresh respectability to subjectivist and irrationalist views about science and related human activities.

1

These days the virulence of the commensurability debate has faded, but it has left its mark on the shape of scientific philosophy and on science itself. Many philosophers currently pay lip service to the 'problem of incommensurability' by insisting on how seriously they take it; only to dismiss it subsequently in a few clipped phrases as being 'settled' and 'solved'. Practitioners of numerous fields, ranging from the natural to the human sciences, from technology to operational research, have been busy unearthing Kuhnian revolutions in areas where they could formerly detect only steady and cumulative progress. For better or worse, one effect of the debate has been to put methodology and philosophy of science on the cultural and scientific map, and sometimes even into the colour supplements and the weekly glossies.

There is of course a danger that, through popularisation and abuse, the term 'incommensurable' will become long in the tooth, and the Kuhnian thesis will lose its original bite. This, to some extent, has happened already. There is a common misconception, spread even to professional methodologists, that almost any kind of conceptual difference is a case of incommensurability, and almost any kind of theoretical change in science is a revolutionary one. This exaggerated use of Kuhnian metaphor does not at all imply, however, that radical relativism is now rampant. Quite the reverse. It indicates only that the sharp language of the radical sixties has melted into the stale metaphor of the conservative eighties. Part of the blame for this lies with Kuhn himself, who has often tried to tone down his views, appease his critics and dissociate himself from the sort of epistemological anarchism that Feyerabend has professed. In some degree this is legitimate defence on Kuhn's part. But anyone who assumes that Kuhn has softened up his rhetoric or conceded substantial ground to the opposition need only reach as far as his contribution to the 1982 meeting of the Philosophy of Science Association, a paper that I shall have more than one occasion to refer to. The message there is plain enough: Kuhn still asserts the disruptive and discontinuous nature of scientific change, and he still acknowledges the full force of the problem of incommensurability.

This 'problem' has many facets and ramifications. In this book I shall be concerned with one aspect of the problem above all others, the aspect which Kuhn himself stresses in the paper mentioned. If pressed to give it a name, I would call it the *logical* problem of incommensurability. It is that aspect of Kuhn's and Feyerabend's doctrine which deals with the cognitive comparability (or noncomparability) of concepts, problems, facts and assertions. On this level, the chief consequence of incommensurability — or, if one takes a different stand, its chief cause — is the impossibility of translating from the language of one scientific theory or conceptual framework into the language of another, rival theory or framework.

Naturally, problems of incommensurability arise at other levels too. There are the problems of comparing cognitive norms, methodological standards, rational goals and cultural values. But, by and large, I shall not be addressing here the pragmatic, historical and ethical dimensions of the incommensurability issue. The reason is simple. I hold that the logical problem still lacks an adequate solution, and that, until progress can be made at this level, there are slim chances to achieve success at any other level.

The various aspects of the problem cannot be divorced from one another. Methodological and cultural norms shape the problems which science investigates as well as the language used for expressing them and the instruments for solving them. But though an inquiry into values and standards may pay dividends in helping to forge commensurability in the wider sense, it cannot be the sole *basis* for understanding science as a rational enterprise. Even if one could show that science makes a judicious choice of its instruments, if there is a rationality gap at the level of concepts and problems then there is a rationality gap in science as a whole because the essential process of *theory* appraisal and choice is, in effect, undermined.

The logical or semantical thesis of incommensurability has come in for rough treatment in the past. Many of its detractors have found the claim itself paradoxical and the arguments for it incoherent. Some have drawn from this the conclusion that there is no case for the traditional, empiricist conception of science to answer. Even among those who on paper have taken the thesis seriously, many have been quick to respond with instant recipes for avoiding incommensurability or its less palatable conse-quences. One should, I believe, view both types of reaction with suspicion. There are self-defeating aspects to most kinds of historical relativism and epistemological scepticism. But only a hardened dogmatist dismisses the relativist or the sceptic as a fool, or listens with one ear like the religious hypocrite.

There have, of course, been more positive efforts to resolve the problem of incommensurability. Questions of meaning, reference, conceptual change and translation have been examined both from the standpoint of general philosophical theories as well as by studying cases from the history of science. In this book some of these approaches are analysed in detail — some in passing, others barely at all. Of those analysed, most are rejected as inadequate, but my criticism attempts to be constructive and, where possible, to build on those features of a rejected account which strike me as defensible. This leads in the end, not to any definite 'solution' to the incommensurability problem, but rather to a series of guidelines, proposals and strategies for reconstructing science as progressive and rational. My own view is that the problem of incommensurability is just not the kind of problem that can be felled in a single blow, and therefore it is senseless to proffer instant remedies for it. I believe the problem can be broached, however, by studying particular episodes of theory change and theory confrontation with the help of the sharpest instruments available, and by developing those methods of logical analysis and intertheory comparison that are best suited to secure a rational picture of the progress of science.

The early chapters of this book are concerned with the work of two writers who have, in different ways, pushed the rationalist conception of science in important new directions over the past decade or so. Wolfgang Stegmüller's 'structuralist' philosophy of science adopts a formal approach to the structure and the dynamics of empirical theories. Based on ideas that originated from Patrick Suppes and Joseph Sneed, this approach now enjoys a wide following, and is associated with what is probably the most ambitious programme of metascientific study to date. I shall also be discussing the work of another contemporary philosopher of science, Larry Laudan. Among those who adopt a more historical perspective in the study of science, Laudan has become one of the most influential and controversial figures. His book *Progress and its Problems* has reawakened interest in the problem-solving aspects of scientific inquiry, and his notion of 'research tradition' has become as central a theme in methodological studies today as Kuhn's 'paradigms' and Lakatos' 'research programmes' were in earlier times.

There are several parallels between Stegmüller's and Laudan's philosophies. Both are instrumentalist in tone and thus swim against the current tide of realism in the philosophy of science. Each of them attempts to bring to bear one especially central, new or revived, idea onto the problem of rational theory appraisal and progress. In so doing, they offer forceful arguments to dispel or to neutralise the destructive effects of the incommensurability thesis. Moreover, they take up Kuhn's and Feyerabend's challenge in a constructive spirit and are prepared, if necessary, to live with the phenomenon of incommensurability without submitting to its negative consequences.

Briefly put, Stegmüller's claim is that difficulties may arise when one tries to present a cognitive comparison of rival scientific theories because one customarily has in mind the wrong notion of 'theory'. By regarding theories as *structures* of a certain sort, rather than as sets of propositions, he argues that one can resolve the main obstacles to theory comparison and accept several of Kuhn's insights into the nature of scientific change, without sliding into an uncontrolled irrationalism. Laudan, who is less concerned to defend Kuhn, maintains that the shared *problems* provide the key cognitive link between rival theories and research traditions in science. But he argues, too, that if problem-solving ability is taken to be the principal indicator of progressiveness, then a rational preference for one theory over its competitors can be justified regardless of any logical or conceptual contact that may or may not exist between them.

These responses to the problem of incommensurability are forthright but, I shall argue, unsuccessful. Each author over-stresses the resources of his own account of scientific progress. Stegmüller takes on the steep task of showing incommensurable theories to be cognitively relatable for (structural) content. Laudan is taxed with the duty to make incommensurable theories rationally comparable, not for content, but for progressiveness. In approaching this issue, each of them leans on his own style of instrumentalism, though neither can be accused of the sort of radical instrumentalism which Feyerabend has characterised with scorn.

Ironically, Laudan and Stegmüller fail in their respective tasks, not because they are instrumentalists, but because they neglect a prime factor in the logical analysis of science which instru-mentalism traditionally assigns a central role: the matter of intertheory *translation*. I shall be arguing repeatedly in these pages that translation is in fact an indispensable ingredient of the process of rational theory comparison. The lack of attention paid to questions of translatability is, I believe, to a large extent responsible for the slow progress made until now in resolving problems of incommensurability and in developing a satisfactory account of the dynamics of science. Moreover, the problem is equally pressing where realist epistemologies and methodologies are concerned; for in my view realists have likewise failed in their efforts to underpin the image of scientific growth as a continuous and well-founded process.

The opening chapter of this book tackles Stegmüller's attempt to defend a Kuhnian rationalism from within the new structuralist metascientific paradigm. His claim that the new perspective reveals an objectivity and rationality in science that the old 'statement view' paradigm conceals is rejected. The reason is that theories which are structurally comparable are also linguistically comparable. In particular, this means that one cannot, as Stegmüller believes, interpret one scientific theory as (structurally) reducible to another and yet conceive that they could be incommensurable at the level of language, i.e. that no translation from one to the other is possible.

I first argued this point a few years ago, and since then Wolfgang Balzer (1985b) has come to the defence of Stegmüller's thesis. His strategy involves a two-pronged counter-assault. One is directed at the premises and assumptions on which my argument is based; the other aims to smother the inference I draw from translatability to commensurability. To lend weight to this second flank, Balzer has proposed an exact explication of incommensurability within the structuralist framework. The upshot of this is that, according to Balzer, not all types of translation, and certainly not the types I envisaged, lead to commensurability. Recently, Wolfgang Stegmüller (1986) has endorsed Balzer's analysis of incommensurability in replying to my criticism of his thesis. My second chapter challenges both sides of Balzer's and Stegmüller's attack. In particular, I defend the inference from reduction to translation, and I also argue that the new structuralist explication of incommensurability is unsound.

Chapter 3 is devoted to Laudan's approach to the problem of incommensurability. Two of his claims are singled out for attention. First, there is his insistence that commensurability is the rule rather than the exception in the history of science because rival theories always share some, solved or unsolved, empirical problems. Secondly, there is his argument that the problem-solving model of science can cope with rational theory appraisal, incommensurability notwithstanding. Laudan's case for the first claim is simply too weak. I suggest that it could be built up from more plausible assumptions and that an analysis of translation is needed in order to lend it proper support. His second claim should also be corrected. He maintains that the problem-solving effectiveness (PSE) of incommensurable theories can be compared. I do not see how. To compare PSEs one has to find a common *scale*, otherwise any numerical similarities or differences of PSE will be entirely fortuitous. But the kind of incommensurability that Laudan is willing (for the sake of the argument) to concede, makes it highly unlikely that such a scale could be devised. It would require, in fact, that one is able to *weigh* the various problems of competing research traditions uniformly. But, if those traditions are incommensurable, they will have not only different characteristic problems, but also dissimilar systems of weights and measures. The result of Laudan's proposal would be like the result of comparing chalk with cheese by determining the price of one and the temperature of the other.

After the predominantly critical emphasis of the first three chapters, the next two take a more positive approach. They explore some of the ways in which a logical analysis of theories and theory change may help to strengthen the rational view of scientific progress. One of the long-standing objections to the application of formal models in the philosophy of science has been that their use of logic tends to be too restrictive and their representation of theories too static. They fail, say the critics, to reconstruct science as a complex, open-ended and evolving product of human reasoning. By characterising comprehensive empirical theories by means of structured networks (of theory elements) and by considering the evolution of such networks over time, Sneed and Stegmüller have clearly shown that part of this objection no longer holds water. But their reaction to the charge that the apparatus of logic is too narrow and limited to reproduce the complexity of empirical science is to drop restrictive systems like first-order logic in favour of no logic whatsoever. On their account, not only are theories to be conceived as structural entities, but relations between theories and the function of theories within a scientific paradigm are likewise to be defined in exclusively structural, mathematical terms. This weakens the structuralists' account of theory dynamics, just as it mars their analysis of commensurability.

Modern developments make it possible, however, to exploit logical tools in metascience without falling prey to the old objection that logic is static and restrictive. In the framework of *abstract* logic, one may treat logic as a *variable* of the system, and recognise that different types of logic and semantics may be appropriate in different contexts and for different theories. In this manner, one can replace the idea of a fixed model-theoretic framework, suitable for all theories, by a flexible system which tailors the model theory to fit the model. Veikko Rantala and I have proposed such a 'general' model-theoretic framework for metascience. In this setting one is able to treat problems of conceptual change and intertheory translation in an open fashion, without having to rely on the sorts of circumlocutions to which structuralists are forced to resort. One can also put to good effect in the study of intertheory relations many of the techniques and results that logicians have established in the fields of non-classical and extended (e.g. infinitary) model theory.

Chapter 4 employs the general model-theoretic approach to

characterise intertheory relations of a broadly 'reductive' kind. I show how different types of theory-reduction and theory-extension can be handled in a uniform manner by means of suitable syntactic and semantic constraints. The resulting framework is then enlarged in Chapter 5 to deal with relations between research traditions and to reconstruct some aspects of theory dynamics. The guiding principle is to represent a research tradition as a structured family of interrelated theories which Rantala and I have called an *ensemble*. Various distinctive features of scientific change can then be more precisely analysed. For example, there is the idea that a research tradition comprises a homogeneous class of theories whose individual and collective evolution is constrained by the background assumptions of the tradition. There is also the idea that scientific progress is to be assessed through the interaction between competing research traditions. Aspects of language and translation come to the fore in reproducing both these features. For instance, the requirement of translatability can be used to express the conceptual uniformity that is internal to a research tradition, as well as the conceptual continuity that holds between its different stages of evolution. Translation *between* research traditions, on the other hand, provides a mechanism for rational appraisal; a means for relocating empirical problems within different conceptual schemes and for explicating the notion of 'anomaly'. An exact reconstruction of these kinds of concepts brings one, I believe, a step closer to Laudan's goal of explaining scientific progress in terms of comparative problem-solving effectiveness.

The last two chapters of this book take up some of the thorny issues of meaning change and reference change which have dominated and still dominate much of the philosophical discussion of incommensurability. Special emphasis is placed on the example of the term 'mass' in the transition from Newtonian to relativistic mechanics; though many of the points made apply to other cases of conceptual change as well. Briefly, my position is as follows. Neither the arguments for meaning invariance nor those for referential stability are compelling. In analysing theory change in science, one cannot therefore take for granted that a concept conserved in name also retains its meaning or its reference. A degree of continuity or overlap of meaning or reference may, in some cases, be established; but this by itself does not ensure that a rational comparison of the theories can be achieved. For this one requires at least some claims made by one of the theories to be translatable into the language of the other. Translatability, however, does not presuppose stability of reference or of meaning; and it may therefore be available even in cases of radical meaning variance. In such cases, a translation must in a suitable sense preserve 'semantic values' like reference, but it need not preserve common words, that is to say, it need not be a homophonic mapping of the expressions shared by two theories.

My contention is that Kuhn's and Feyerabend's arguments for meaning variance do not entail without further ado the thesis of untranslatability. To the extent that their arguments are correct, however, they do on occasion provide valuable criteria for the adequacy of a translation; they suggest certain properties that translations ought to fulfil. A successful response to the challenge of the doctrine of incommensurability thus consists in examining concrete examples of theory change in science and providing translations of the right kind which apply there. In the last chapter, some efforts in this direction are made.

The case of classical mechanics and special relativity affords a fine example of this. The two theories share common terms, like 'mass', which, however, arguably undergoes a shift in meaning from one theory to the other. Nevertheless, translation from the language of the earlier theory into the language of its successor is not precluded. In fact, a suitable approach to translation has been opened up by Veikko Rantala's (1979) insight that certain logical methods, especially the tools of nonstandard analysis, can be profitably applied to give a rigorous characterisation of the 'correspondence relation' that holds between classical and relativistic mechanics, CM and RM. Making use of the more complete reconstruction provided in Pearce and Rantala (1984a), in Chapter 7 I outline the way in which a recursive translation from the language of CM into the language of RM can be defined. I also argue that this translation satisfies the kinds of adequacy constraints which Kuhn and Feyerabend have invoked in their discussions of this example.

I maintain that, at least for one important type of scientific change (of which the replacement of CM by RM is perhaps the most celebrated example), this analysis offers a new perspective on the problem of incommensurability. What it shows, in short, is that one may uphold a view of meaning variance similar to Kuhn's, as well as his assertion that no 'theory-neutral' language is available for expressing the rival claims of CM and RM, without inferring that no ordinary logical comparison of the two theories can be made. For the language of RM is already rich enough to reproduce (under nonhomophonic translation) the central laws of Newtonian mechanics; and thus a rational comparison of the two theories within the relativistic framework is feasible.

To my knowledge, this conclusion differs in several important respects both from the influential picture of scientific change which Kuhn has drawn as well as from the multitude of conflicting accounts which his critics have offered us over the years. Since science as a whole is so rich in concepts and so diversified in its patterns of conceptual change and development, one cannot expect that methods of translation applicable in one context will automatically be adequate in another. One cannot, therefore, extrapolate too far from individual case studies, however pertinent, to make global inferences about the nature of scientific change. Even the most mature theories of scientific progress are still, it seems to me, in their infancy. At the same time, though the analysis I propose here does not deliver any fully-fledged theory of scientific change and rationality, I hope that it will provide some comfort for the rationalist picture of science and some materials for future study. I also think one can take an optimistic view about the long-term prospects for resolving the central philosophical problems about the nature of scientific change. There is an invaluable feedback from science to metascience and methodology. As science itself grows, so our resources for understanding science, reasoning about it and rationally reconstructing it, grow too. Evidently there is no room for complacency here: the philosophy of science must work hard to keep pace and move with the times!

STEGMÜLLER ON KUHN
AND INCOMMENSURABILITY

The so-called *incommensurability thesis* has emerged as undoubt-edly the single most controversial and questioned feature of Thomas Kuhn's *The Structure of Scientific Revolutions*. Many critics, unable to accept the proposition that rival pairs of theories or paradigms might be noncomparable by any of the usual standards, accused Kuhn of thus propagating a subjectivist, rela-tivist and irrationalist model of scientific change. It is well known that Kuhn himself has steadfastly denied this charge. He maintains that the development of science, even in its revolutionary phases, can be rationally appraised and understood; but only by means of a broader conception of rationality than philosophers and historians of science had hitherto conceived. The ensuing battle between Kuhn and his critics has lasted more than twenty years, with little sign of any truce in the offing. And though Kuhn has clarified, modified and even abandoned some of his earlier views, he remains today as staunchly as ever committed to the thesis of incommensurability, as a recent attack on his critics testifies (Kuhn, 1983).

'Comparability' tends, of course, to be an ill-defined and woolly notion. Whether two scientific theories or paradigms can be cognitively connected depends much on the form in which they are presented as well as on the standards of comparison operative. It is thus easy to conceive a pair of theories as logically related under one conceptual scheme, incommensurable with respect to another. Such ambiguity in the notion of comparability can, it seems, be exploited to the advantage of those seeking to further the Kuhnian theory of scientific progress whilst denying the charges of extreme relativism and irrationalism. It has been most adroitly employed to this effect by Wolfgang Stegmüller, who has offered a detailed and vigorous defence of the central tenets of the Kuhnian theory.

The nub of Stegmüller's argument rests in the claim that the phenomenon of incommensurability flows from a common but inadequate conception of a scientific *theory* as a set of *propositions*. Rival but conceptually different theories may, according to Stegmüller, be incommensurable because, viewed as classes of sentences, they are not logically relatable. However, when correctly construed, not as statements but as *structures* of a certain sort, the problem of incommensurable theories is resolved: structures can be canonically related where sentences cannot. Thus, he claims, it is possible to accept the presence of incommensurable theories in science without thereby inferring the noncomparability and nonreducibility of competing paradigms; that is, without supposing that theory appraisal and choice is an irrational or noncognitive affair. In this manner, by switching to what he calls a *structuralist* (or nonstatement) view of theories, Stegmüller aims to salvage a substantial portion of Kuhn's account of science from the charges of subjectivism, relativism and irrationalism.

Stegmüller's approach promises a radical solution to the problem of incommensurability, since it demands a thorough revision of our basic metascientific categories, such as 'theory', 'paradigm', 'reduction', and the like. In so doing, it also compels changes in Kuhn's own picture of scientific growth; not least it expounds the merits of formal tools that Kuhn himself eschews. But, whilst the structuralist framework as such remains to a large extent independent of specific scientific methodologies, Stegmüller's proposed reconstruction of theory dynamics has nonetheless a quintessentially Kuhnian flavour.

There are several reasons to take a special interest in

Stegmüller's central assertion. In the first place, it provides a substantial theoretical argument to back Kuhn's 'rationality' claim against the protestations of the critics. On these grounds alone it merits careful consideration. In addition, if correct, Stegmüller's contention would lend indirect but weightly support to the structuralist view of theories. For, if that approach is able to yield a persuasive account of scientific development as a rational process where other modes of reconstruction fail, this would amount to a strong case in its favour. Moreover, it opens the way to a re-evaluation of Kuhn's account of science. Kuhn's views have exerted an enormous influence on scientists and methodologists alike; yet many have stopped short of a wholehearted endorsement of Kuhnian theory largely as a result of the extreme relativism which the incommensurability thesis seems to imply; and this is exactly the issue to which Stegmüller's arguments are primarily addressed. Lastly, a proper investigation of Stegmüller's claim — and later developments of it to be discussed in the next chapter — obliges one to treat several key and independently significant issues in the logic of theory-change; particularly questions concerning the connections between intertheory translation, reduction and commensurability.

Despite its importance, Stegmüller's central thesis on commensurability has received less attention than it deserves. Most commentators on the structuralist approach have dwelt on other features of the Kuhn-reconstruction, or on the logical basis of the nonstatement view of theories.[1]* And the impact of the thesis on the debate between Kuhn and his critics has been only marginal, notwithstanding Kuhn's own quite favourable response to it. In some cases at least, this is a consequence of a superficial and inadequate appreciation of the argument. To mention one example, Siegel (1980), who offers a quite thorough survey of the

---

* Notes are given at the end of the text.

problem of comparability and commensurability in the light of Kuhn's recent writings, and who sides with Scheffler in the accusation of subjectivism and irrationalism, brushes aside Stegmüller's thesis without producing a single shred of evidence against it.

As far as one can tell, Siegel has simply misunderstood the matter, however. He refers to a number of passages from Stegmüller (1973/76) to support his claim that Stegmüller fails to answer Kuhn's critics because his reconstruction does not incorporate the irrationalist elements in Kuhn's philosophy. But Stegmüller's whole point is just to show that a 'reasonable' (i.e. suitably reformed and reformulated) Kuhnian is not committed to irrationalism at all; his reconstrual is thus rationalist *by design*. Moreover, Siegel cites Stegmüller's claims out of context. For instance, he correctly observes that according to Stegmüller (1973/76) "competing paradigms are not incommensurable", and "old paradigms are reducible to new ones". Quite so; but Siegel neglects to mention that here Stegmüller has already switched to the new metascientific 'paradigm' of the structuralist view according to which a superseded theory *is* generally reducible to its successor, but under a quite different conception of reduction than that used by Kuhn and his critics. Perhaps Stegmüller's use of the term 'incommensurability' here is somewhat misleading, but any confusion that arises is immediately resolved in a later work (1975) to which Siegel also refers. There, in particular, Stegmüller summarises his conception of a scientific revolution in the following passage which I shall denote by *SR*:

> A 'proper scientific revolution' exhibiting scientific progress consists in a theory T_1 being supplanted by a theory T_2 whereby (1) T_1 and T_2 are incommensurable and (2) T_1 is reducible to T_2 but not vice versa.
>
> (1975, p. 198)

Stegmüller goes on to make it plain that in this remark 'reduction' is to be understood within the structuralist conception, whilst 'incommensurability' is to be read as the absence of logical relations between theories construed as classes of statements. There is apparently no conflict between the two sub-claims of *SR* because the notion of 'theory' embodied in clause (1) is quite different from that intended in clause (2).[2]

SR represents the crux of Stegmüller's 'defence' of Kuhn from his critics, for it maintains exactly what most of the critics have denied: the possibility of rational and objective comparison of incommensurable theories. For the remainder of this chapter I want to examine *SR* in some detail, especially the structuralist notions of 'theory' and 'reduction' underlying it. I shall argue, in fact, that *SR* cannot work as a correct characterisation of scientific revolutions and that, accordingly, this central feature of Stegmüller's 'rationality' thesis is mistaken.

To this end, it will not be necessary to evaluate the adequacy of Stegmüller's concept of reduction in its entirety. For, independent of the question whether clause (2) of *SR* actually holds in typical cases, it can, I believe, be shown that no pair of theories T_1, T_2 can jointly satisfy (1) and (2). The reason is simply that if a theory T_1 is reducible in the structuralist sense to a theory T_2 then, under plausible assumptions, it can be shown that the language of T_1 is translatable into the language of T_2. In short, structurally comparable theories are linguistically comparable too. And this, I shall claim, ought to be quite sufficient to ensure their commensurability.

1. THE STRUCTURALIST VIEW OF THEORIES

In order better to grasp the full import of Stegmüller's thesis (*SR*) one should first consider some basic properties of the structuralist conception of theories and intertheory relations. For present purposes a quite brief and sketchy survey will suffice.

The structuralist view has its origins in an approach, popularised by Suppes, towards axiomatising physical theories by the method of defining a set-theoretic predicate. It is well known from mathematics that there are usually quite distinct ways in which a given concept can be formally presented. Many algebraic structures, like groups, rings, fields and so forth, can be delineated by means of a set of sentences in a first-order language: the structures in question are simply the *models* of the first-order axioms, in the usual Tarskian sense. Alternatively, it is more common in ordinary textbooks of algebra directly to define a *group*, say, to be a structure consisting of a nonempty set X together with a binary relation ' $\cdot$ ' on X, called 'product', and then simply list the properties that the structure $\langle X, \cdot \rangle$ is required to fulfil. In so doing, one essentially specifies a set-theoretic predicate, 'is a group', satisfied by all and only those objects that one designates by the term 'group'. Suppes and his collaborators were perhaps the first explicitly to recommend the second approach as a method for characterising physical theories in general.

This idea underwent further elaboration in the work of Adams (1959) and later Sneed (1971). In order to reflect the fact that a physical theory is not generally a system of pure mathematics, but is an object that can be empirically applied, Adams proposed to characterise a theory T as an ordered pair $T = \langle M, I \rangle$, where M is the class of the theory's *models* (structures satisfying the set-theoretic predicate), and I is a collection of structures representing the actual physical situations (systems, states, processes, or whatever) to which the theory is intended to apply.

Sneed found it useful to introduce further distinctions here. The intended applications of T, he claimed, should be structures appropriate not to the full conceptual apparatus of the theory but only a part of it — namely, to those physical concepts and quantities that can be measured (in a suitable sense) without recourse to the theory T itself. To make this idea precise, he proposed a general partition of terms into *T-theoretical* and *T-*

nontheoretical ones, for any given theory *T*. The full conceptual structure of the theory can then be represented by a class *Mp* of so-called *potential models* each element of which contains a function or relation corresponding to each term (theoretical or nontheoretical) of *T*. Letting *r* be the function which removes from each potential model 𝔐 all components that correspond to *T*-theoretical terms, Sneed then designated the collection *Mpp* (= *r*[*Mp*]) the class of *partial potential models* of *T*, its elements representing potential applications of the theory. These collections are specified in such a way that the class of models *M* of *T* is a subclass of *Mp*, and the class *I* of intended applications of *T* is a subclass of *Mpp*.

As an additional refinement, Sneed introduced the notion of a *constraint* *C* for a theory *T*, being defined as a subcollection of 𝒫(*Mp*) satisfying certain properties. On need not be too concerned here with the details of this notion, though intuitively a constraint may be thought to express higher-order conditions on the models of the theory.

In summary, then, a theory is described in structuralist terms by a mathematical formalism or *core* $K = \langle Mp, Mpp, M, r, C \rangle$ together with a class $I \subseteq Mpp$ of intended applications. This, to all intents and purposes, is the theory-concept which Stegmüller elaborates and which underlies his claim *SR*. Perhaps its most striking feature is the absence of any logical or syntactic constituents: a scientific theory is regarded as an exclusively structural entity. In particular, any reference to the vocabulary or primitive terms of the theory occurs only obliquely, in what Stegmüller calls its *presystematic* (i.e. unreconstructed) form; and is completely dropped from the *systematic* or reconstructed version. Likewise, the theory's laws, usually construed syntactically, are here captured by the class of models, *M*, whilst other relevant features of a lawlike nature may be represented in the constraint, *C*.

What does it mean in this framework to say that a theory $T = \langle\langle Mp, Mpp, M, r, C\rangle, I\rangle$ is reducible to a theory $T' = \langle\langle M'p, M'pp, M', r', C'\rangle, I'\rangle$? Evidently, appropriate connections have to be specified between the various components. M, M'; I, I'; etc. Suppes' own answer to this question takes the following form (cf. Suppes, 1957). If T is reducible to T', then, given any model $\mathfrak{M}$ of T, it must be possible to find a model $\mathfrak{M}'$ of T' from which one can construct a model $\mathfrak{M}^*$ that is isomorphic to $\mathfrak{M}$. Though Suppes does not make the notion of 'construction' here fully explicit, one may plausibly conjecture that he means $\mathfrak{M}^*$ to be in a suitable sense *definable* from $\mathfrak{M}'$. In other words, given $\mathfrak{M}'$ and appropriate defining conditions, one can construct, up to isomorphism, the original model $\mathfrak{M}$. A standard construction of this kind would be the Cartesian interpretation of Euclidean geometry which reduces this theory to the arithmetic of the reals. Another would be the method whereby a model of the rationals is obtained from a model of the integers by forming Cartesian products and factorising by definable equivalence classes. If the theories and defining conditions involved are first order, then Suppes' criterion for reduction is essentially a semantic variant of the usual syntactic notion of (*relative*) *interpretability* between elementary theories. Loosely speaking, T is interpretable in T' if there is a recursive translation which maps T-theorems to T'-theorems. Its semantic counterpart is an operation which takes models of T' to models of T; and Suppes' requirement then states that *all* T-models are in this manner retrievable (up to isomorphism) from models to T'.

Adams' (1959) approach to reduction incorporates a refinement and also a generalisation of Suppes' criterion. He requires that the relation between the models of the two theories induces a similar connection between their intended applications, so that every member of I has a suitable counterpart in I'. But Adams makes no demands on *how* models of T are to be obtained from models of T'. Thus, for him, reduction is fully described simply by a *relation* $\rho \subseteq M' \times M$. This idea is retained in subsequent developments of the structuralist programme, e.g. in Sneed (1971), Stegmüller (1973/76) and Mayr (1976). In these works one loses sight of the fact that ordinarily if $\rho(\mathfrak{M}', \mathfrak{M})$, then $\mathfrak{M}$ should in some sense be constructible or definable from $\mathfrak{M}'$. In Sneed's formulation (almost exactly taken over by Stegmüller), the reduction of T of T' is basically characterised by a (partial) function $\rho: M'p \rightarrow Mp$, with the property

(0)   for all $\mathfrak{M} \in \text{Dom}(\rho)$, $\mathfrak{M} \in M' \Rightarrow \rho(\mathfrak{M}) \in M.$[3]

Various additional conditions require ρ to induce canonical relations between Mpp and $M'pp$, I and I', and between C and C'. I shall not enter into details here, except to point out that each of these extra desiderata is designed to reflect the fact that the reducing theory should be the more 'comprehensive' of the two and that the entire cognitive and empirical 'content' of the reduced theory should be absorbed by it. Notice once again that the map ρ and each of its formal adequacy conditions are described in purely set-theoretic jargon; no linguistic or logical concepts are involved.

Like Kuhn, Lakatos and Laudan, structuralists also consider larger units of methodological appraisal than single theories. They define a theory *net N* to be a tree-like array of theories, partially ordered by a relation called *specialisation* which links theories possessing the same conceptual structure (potential models). Roughly, a theory T_1 is a specialisation of a theory T if it has additional ('special') laws and constraints that operate on particular applications of T. A theory net N may be taken to represent the set of theories in a given field of science which the scientific community accepts at time t. One can then obtain a diachronic analysis of theories by indicating a so-called theory *evolution* $E = \langle \mathbf{N}, \leqslant \rangle$, where $\mathbf{N}$ is a sequence of nets and $\leqslant$ is

an historical ordering. If t and t' are time intervals, with t earlier
than t' ($t \leqslant t'$), then the scientific progress attained in the
transition from t to t' is assessed by analysing the formal relations
between the nets N_t and $N_{t'}$ belonging to **N** which the scientific
community holds at those respective times. A theory evolution E
is said to be *Kuhnian* when there exists a *paradigm* theory $\langle K, I \rangle$
such that for every net $N \in$ **N** and every theory $T' = \langle K', I' \rangle$ of
N, K' is a specialisation of K, and I' contains at least some
elements of I. Moreover, the scientific community must recognise
these elements of I to be *paradigm* applications.

   The relation of reduction is generalised so as to be applicable
to both nets and evolutions. In each case reduction is a canonical
correlation of these 'larger' structures, defined in terms of the
individual reductions that obtain between their component theories. Whereas normal science in Kuhn's sense is characterised by
a single (Kuhnian) theory evolution E, a revolutionary change
of theory or paradigm is represented by the transition from an
evolution E to an evolution E' which possesses a conceptually
different paradigm theory. E' may mark a progressive (though
noncumulative) development in this field of science if the earlier
evolution E is reducible to it.

   With this type of analysis of theory change and progress,
Stegmüller aims to reproduce the main ingredients of Kuhn's
account of science — including potential incommensurability at
the linguistic level — whilst rebutting the allegations of relativism
and irrationalism.[4]

2. AN ANALYSIS OF THE STRUCTURALIST
CONCEPT OF REDUCTION

Stegmüller lays considerable stress on the fact that the nonstatement view of theories is to be reckoned as a radical alternative to
the traditional conceptions employed by Kuhn and his critics in

their debates over rationality and commensurability. He is equally single-minded in his insistence that the new model of intertheory reduction, central to the rationality thesis *SR*, is superior to the customary account under which reduction is construed as a special kind of deductive relation between theories. As he puts it:

> For the 'statement view', reduction problems can only turn on one thing: namely, the 'inference relations between classes of sentences'. *However, instead of arguing from noninferability to nonreducibility, we will argue that an adequate concept of reduction cannot be defined in terms of inference.* (1976; p. 216).

Now that we have a rough outline of the new metascientific paradigm, we can begin to assess the extent to which it might provide theoretical support for the Kuhnian theory of science; and whether, in fact, the claim of 'incommensurability without irrationality' can be sustained.

First, notice that Stegmüller's thesis is vulnerable to criticism on a number of grounds. One can, and with good reason, attack his contention that the structuralist framework yields a richer and more adequate conception of science than any 'statement view' alternative. One might also question the adequacy of Sneed's explication of reduction, even from a structuralist perspective. And a strong case can be made that there are some aspects of the proposed 'Kuhn-reconstruction' that are unsatisfactory for the purposes of achieving a convincing account of theory development in science. All these arguments have been raised on different occasions and by different authors in response to the structuralist philosophy of science.[5] For the most part, however, they do not reach the roots of the issue I am at present concerned with, because they do not settle one way or the other the question whether a claim like *SR* can plausibly be defended. Such arguments might, therefore, lead to modifications and improvements in the structuralist approach without affecting the viability of

Stegmüller's central philosophical claims; indeed his position might ultimately be enhanced by this kind of criticism.

I want to suggest, however, a more direct route towards evaluating Stegmüller's main thesis. As it stands, the argument for *SR* is incomplete at one vital point. In order to defend *SR* it is necessary (though not of course sufficient) to show that at least in principle a pair of scientific theories T and T' connected by a structuralist reduction relation *could* be incommensurable; in other words that no relation of translation and deduction holds between T and T' when they are *re*construed as classes of statements. In short, it has to be demonstrated that the new model of reduction is *de facto* and not merely *prima facie* noninferential. If such an argument could be supplied, and the new model of reduction shown to be adequate on independent grounds, then Stegmüller's thesis would form the basis of a highly promising defence of Kuhn. If, as I believe, the argument cannot be given because a structuralist reduction relation actually establishes the commensurability of the theories it connects, then *SR* collapses. In that case Stegmüller's defence of Kuhn is robbed of its only really substantial theoretical weapon. In the absence of an assertion like *SR*, Stegmüller's reconstruction of Kuhn is at best of exegetical interest. It may clarify and even improve on the Kuhnian theory of scientific progress (and remain methodologically valuable for that), but it cannot add weight to *justify* the Kuhnian view or contribute decisively to the 'rationality' debate between Kuhn and his critics.

Before we embark on a proper dissection of Sneed's and Stegmüller's reduction concept, let me mention some earlier attempts to compare its ancestor, the Suppes—Adams concept, with more traditional approaches to the subject. In one, much-cited study of reduction, Schaffner (1967) considers Suppes' version of reduction and finds it to be a weakened variant of the standard account (which he attributes mainly to Nagel and Quine) according to which the reducing theory, together with auxiliary assumptions, bridge laws, etc., is supposed logically to entail the reduced theory. On the other hand, Eberle (1971) analyses Adams' notion of reduction and reaches the rather different conclusion that it is a special case of the standard concept, identified by Eberle with relative interpretability in the sense already mentioned.

Unfortunately, neither of these two accounts can be relied upon for our present aims. In particular, Schaffner's remarks are based upon an incorrect claim (the 'Theorem', p. 145 of his 1967).[6] And Eberle's conclusion, though technically sound, holds under a very strong assumption, namely that the structural reduction relation is defined by means of an antecedently available translation connecting the languages of the two theories.[7] The possible existence of such a translation is, however, exactly what is at issue here, and we cannot therefore take it for granted in our analysis of structuralist reduction. In addition, both Schaffner's and Eberle's discussions are confined to first-order theories, whereas one of the chief arguments used by Sneed and Stegmüller in favour of their account is that it permits one to relax such restrictive assumptions and study relations between 'real scientific theories' of whatever logical complexity.

Can we do justice to this latter aspect of the structuralists' approach and still obtain a reasonable rendering of what their reduction concept amounts to when the linguistic dimension of scientific inquiry is re-introduced? I think we can, and I now want to sketch the promised argument to show that if a theory T is reducible in Stegmüller's sense to a theory T', then T and T' are commensurable. The kernel of the argument is in three parts. The first step consists in adopting several conventions and technical assumptions under which a structuralist reduction relation is or induces a certain logically definable relation on classes of models. Step two invokes some metalogical results in the shape of *uniform reduction* theorems which establish that the definable relations in question imply a translatability relation between the languages of the theories under scrutiny. The third step is the inference from translation and deduction to commensurability. The second step rests on purely formal results and should be to that extent uncontroversial, given the assumptions of step one. The other parts of the argument are of course open to criticism and have in fact been questioned by Wolfgang Balzer (1985a; 1985b) in a new defence of *SR*. I shall postpone to the next chapter a discussion of Balzer's objections.

To get the argument underway, let us suppose that T and T' are theories whose classes of (potential) models are (Mp and $M'p$) M and M' respectively. Let ρ be a partial function from $M'p$ into Mp by which T is reduced to T' in Sneed's sense; that is to say, condition (0) holds. Naturally, this is a highly simplified characterisation reflecting only a portion of what is involved in the structuralist account. Nevertheless it will suffice for the present, since I want to draw out only the most basic consequences of the reduction relation needed for the argument.[8]

Next we require four technical assumptions which I shall first list and then discuss in more detail. They are as follows:

(1)   Elements of Mp (respectively $M'p$) can be regarded as structures or models of a fixed, many-sorted similarity type to be denoted by τ (respectively τ'). Equivalently: $Mp \subseteq \mathrm{Str}(\tau)$ and $M'p \subseteq \mathrm{Str}(\tau')$, where for any type τ, $\mathrm{Str}(\tau)$ denotes the collection of all structures (models) of type τ.

(2) ρ is or induces a single-valued algebraic (isomorphism-preserving) relation on structures, such that the relation can be projectively defined in a regular logic L.

(3)   The types τ and τ' are disjoint.

(4) L has the Interpolation Property; i.e. a generalised version of Craig's Interpolation Theorem holds for the logic L.

In assumption (1), by 'similarity type' is meant a (many-sorted) vocabulary or signature in the usual sense. Thus τ can be thought of as a collection of symbols (e.g. predicate letters) and sorts, and a structure of type τ is, in the ordinary way, a function that assigns a nonempty domain to each sort of τ and a relation or function (over the appropriate domain(s)) to each symbol of τ. Relations and functions may be of mixed sort, and individual constants can be incorporated as 0-ary functions. As far as the structuralist metatheory is concerned, (1) appears to be a plausible and unrestrictive assumption, for it merely makes explicit the fact that the potential models of a theory are intended to be structures 'of the same kind'. It is virtually inconceivable that the potential models of an actual scientific theory could not be represented by a class of structures of some given type in this sense.

The second assumption requires somewhat more explanation. First, a relation $\rho \subseteq \mathrm{Str}(\tau') \times \mathrm{Str}(\tau)$ is said to be *algebraic* if for any $(\mathfrak{M}', \mathfrak{M}) \in \rho$, and $(\mathfrak{N}', \mathfrak{N})$ isomorphic to $(\mathfrak{M}', \mathfrak{M})$, $(\mathfrak{N}', \mathfrak{N}) \in \rho$. In particular, that ρ is *single-valued* means that it comprises pairs $(\mathfrak{M}', \mathfrak{M})$, where $\mathfrak{M}' \in \mathrm{Str}(\tau')$, $\mathfrak{M} \in \mathrm{Str}(\tau)$, such that for any $\mathfrak{M}', \mathfrak{M}, \mathfrak{N}$, if $\rho(\mathfrak{M}', \mathfrak{M})$ and $\rho(\mathfrak{M}', \mathfrak{N})$, then the identity mapping of $\mathfrak{M}'$ can be extended to an isomorphism of $(\mathfrak{M}', \mathfrak{M})$ onto $(\mathfrak{M}', \mathfrak{N})$.[9] To say that ρ is *projectively definable* in a logic L means that there is a similarity type $\tau^* \supseteq (\tau' \cup \tau)$ and a set of L-sentences S in the type τ^* such that

$$\rho(\mathfrak{M}', \mathfrak{M}) \quad \text{iff} \quad \exists \mathbf{D}, \exists \mathbf{R}, (\mathfrak{M}', \mathfrak{M}, \mathbf{D}, \mathbf{R}) \vDash_L S,$$

where $\vDash_L$ is the truth relation for L and $\mathbf{D}, \mathbf{R}$ are respectively sets of domains and relations corresponding to sorts and symbols in $\tau^* - (\tau' \cup \tau)$. Lastly, I suppose that the logic L satisfies some standard closure properties, abbreviated by the label 'regular'.[10] I shall not further discuss the notion of 'logic' here, except to mention that it is interpreted in an extremely general sense to encompass both first-order logic and its usual extensions (e.g. infinitary logics and logics with added quantifiers) as well as many 'nonclassical' logics (e.g. modal and many-valued logics).[11]

Since this notion of logic is so broadly construed, the supposition contained in (2) that the reduction relation ρ is appropriately definable in some logic L can scarcely be called into question. However, the qualification "or induces" in (2) is important, because generally speaking the single-valuedness of ρ is weaker than Sneed's requirement that a model of T is uniquely obtained from a model of T' (i.e. that ρ is a function). Whilst, by employing an inclusive metalanguage, we can often pick from all isomorphic copies a unique representative, if ρ is to be definable in the sense of (2) then we should generally suppose that it determines models of T up to isomorphism, as Suppes' criterion of reducibility demands. Let us assume, therefore, that a Sneed reduction can be equivalently described by such a single-valued relation ρ, in the sense that the 'intuitive content' of the reduction is not altered when we represent it by a relation of this kind.[12]

Condition (3) requires T and T' to possess disjoint vocabularies (and sorts). As will be seen later, this constraint can be lifted at a price. However, the assumption itself is not genuinely restrictive at all, because one can always replace one of the theories by an alphabetical variant simply by changing the names of all its primitive terms. This formal device is available in the present framework, since one of our assumptions is that the logic L admits name-changes.[13]

The last condition (4) expresses a limitation on the type of logic that is allowed to define the reduction. In single-sorted, first-order logic Interpolation is just the property described by the well-known theorem of Craig (1957). In a suitably generalised form it is one of the many properties that logicians have studied when investigating extensions of, and possible alternatives to, elementary logic.[14] Generally, Interpolation — like compactness and axiomatisability — is considered to be a desirable property for a logic to possess. The fact that it fails to hold for a number of important extensions of first-order logic has even led to a study of conditions under which a logic without Interpolation can be strengthened (closed) to a logic with this property. To mention just a few examples, Interpolation holds for intuitionistic and some standard modal logics, for admissible fragments $L_{\mathscr{A}}$ of $L_{\infty\omega}$ (where all members of $\mathscr{A}$ are hereditarily countable), and for $L_{\omega_1\omega}$. It fails, however, for 'large' infinitary logics, $L_{\kappa\kappa}$, for second-order logic, and for some logics with added quantifiers, like $L_{\omega\omega}(Q_0)$ and $L_{\omega\omega}(Q_1)$; though the so-called Δ-*closures* of weak second-order logic, L_w, and $L_{\omega\omega}(Q_0)$ have Interpolation.[15]

(4), when coupled with (2), is the key assumption on which the whole argument rests. Its reasonableness can be defended, I think, by observing that there is a relatively rich variety of logics that enjoy the requisite property, both of the sort which extend the expressive power of first-order logic, as well as those which deviate in other respects from standard quantification theory. This should give the would-be reductionist ample scope to characterise intertheory relations in the manner indicated. Notice that I am not suggesting that all examples of reduction in science can or must be reconstructible on the basis of assumptions (1)—(4). Neither do I claim that Sneed's explication of reduction is necessarily adequate in this respect. My claim is rather that a substantial class of Sneed reductions can be equivalently described by a relation satisfying (1)—(4), plus of course the further conditions that I shall not discuss here.

We are now ready to proceed to the second stage of the argument which consist in applying what are known as uniform reduction theorems in order to infer a syntactic relation between T and T', given the stated model-theoretic relation. As the name suggests, such theorems as a rule provide conditions under which a specified relation or operation on classes of models allows for the reduction of properties of models in the range of the relation (in this instance: structures in Mp) to properties of models in the domain (structures in $M'p$). The properties in question are expressed by sentences in the appropriate languages (types) and the uniform reduction then associates sentences of the one language with sentences of the other. One can regard this procedure as determining a *translation*; in our case a translation from the language of the reduced theory into the language of the reducing theory.

The assumptions I have listed are by no means the only possible hypotheses under which uniform reduction theorems can be applied to infer translatability phenomena given a situation of reduction in the structuralist sense. In the next chapter I shall mention some alternative choices. The advantage of these particular conditions is that they are weak and therefore plausible. They 'add' very little to what is already contained, implicitly or overtly, in Sneed's concept. And yet they suffice for applying a very general kind of uniform reduction theorem, due to Feferman (1974). As will shortly become apparent, the theorem is *so* general that it yields virtually no lexical information about the kind of translation that Sneed's relation induces; this is the price we have to pay for profiting from a widely applicable result. However, the reader should bear in mind too that our present goal is also quite modest: that of showing that at least *some* kind of deductive connection obtains between Sneed-reducible theories.

From now on let τ and τ' be disjoint types and ρ an algebraic relation, $\rho \subseteq \mathrm{Str}(\tau') \times \mathrm{Str}(\tau)$. Denote the collection of all L-sentences of type τ (respectively τ') by $\mathrm{Sent}_L(\tau)$ (respectively $\mathrm{Sent}_L(\tau')$). For any $\varphi \in \mathrm{Sent}_L(\tau)$, I shall say that a given $\psi \in \mathrm{Sent}_L(\tau')$ is a *translation* of φ, relative to ρ, if for all $\mathfrak{M}'$, $\mathfrak{M}$,

$$\rho(\mathfrak{M}', \mathfrak{M}) \Rightarrow (\mathfrak{M}' \vDash_L \psi \Leftrightarrow \mathfrak{M} \vDash_L \varphi).$$

Further, I shall say that ρ determines a translation of $\mathrm{Sent}_L(\tau)$ into $\mathrm{Sent}_L(\tau')$, providing every $\varphi \in \mathrm{Sent}_L(\tau)$ has a translation ψ

$\in \text{Sent}_L(\tau')$, relative to ρ. Feferman (1974) proves the following result:

Uniform Reduction Theorem. *Suppose that L has the Interpolation Property and that ρ is projectively definable in L. For any $\varphi \in \text{Sent}_L(\tau)$, if*

(i) $\forall \mathfrak{M}', \mathfrak{M}, \mathfrak{N}, \quad \rho(\mathfrak{M}', \mathfrak{M}) \,\&\, \rho(\mathfrak{M}', \mathfrak{N}) \Rightarrow (\mathfrak{M} \vDash_L \varphi \Leftrightarrow \mathfrak{N} \vDash_L \varphi)$, *then φ has a translation $\psi \in \text{Sent}_L(\tau')$, relative to ρ.*

As an immediate corollary we see that if ρ is a reduction of T to T' satisfying (1)–(4), then ρ determines a translation of $\text{Sent}_L(\tau)$ into $\text{Sent}_L(\tau')$; for the fact that ρ is single-valued here ensures that (i) holds for all sentences φ of $L(\tau)$.

Translation is here conceived in the widest possible terms to be a correlation of sentences with sentences uniformly determined by the model-theoretic correspondence. The upshot of the argument thus far is that there exists a translation *function*, Γ say, where $\Gamma: \text{Sent}_L(\tau) \to \text{Sent}_L(\tau')$, that *respects* ρ in the sense that

(5) $\forall \varphi \in \text{Sent}_L(\tau), \quad \forall \mathfrak{M}' \in \text{Dom}(\rho),$
       $\mathfrak{M}' \vDash_L \Gamma(\varphi) \Leftrightarrow \rho(\mathfrak{M}') \vDash_L \varphi;$

where ρ is now treated as a single-valued operation and $\Gamma(\varphi)$ is a translation of φ, relative to ρ. If S is the set of sentences in the expanded type $\tau^* \supseteq (\tau' \cup \tau)$ which 'defines' ρ in accordance with (2), then we have

(6) $\forall \mathfrak{M}^* \in \text{Str}(\tau^*), \quad \mathfrak{M}^* \vDash S \Rightarrow$
       $(\mathfrak{M}^* \vDash \varphi \Leftrightarrow \mathfrak{M}^* \vDash \Gamma(\varphi)).$

This clarifies the sense in which a sentence and its translation are 'equivalent' under ρ. Clearly, if L has a suitable bi-conditional connective '$\leftrightarrow$', (6) can be recast in the form

(7) $S \vDash (\varphi \leftrightarrow I(\varphi)).$

And if L has a complete set of proof procedures, this expression can be further improved to

(8) $S \vdash (\varphi \leftrightarrow I(\varphi))$.

in the sense of syntactic consequence.

In general, ρ cannot be taken to determine the translation Γ *uniquely*, nor is there any guarantee that this syntactic correlation is an effective or recursively definable mapping. All the same, the presence of this translation, with the properties indicated, seems to be an almost inescapable consequence of the structuralist definition of reduction. It provides ample grounds on which to claim that, contrary to the spirit of SR, a relation of reduction cannot obtain between theories that are in any reasonable sense incommensurable.

3. FURTHER CONSEQUENCES

The foregoing argument does not yet quite deliver all that it was billed to. I also promised to show that a Sneed reduction of one theory to another establishes a deductive relation between them; and there is an important part of this claim that still remains to be verified.

In a nutshell, the argument as it stands says roughly (but only roughly) this: if it is possible, for any Sneed reduction of T to T', to find an equivalent linguistic characterisation of the relation, then, under suitable assumptions, one may infer the existence of a translation from the language of one theory into the language of the other. 'Linguistic characterisation' here, means, of course, that ρ can be 'defined' by a set of sentences S in the manner sketched above. The assumption is generous, because there is no restriction on the type τ^* (other than the trivial constraint that both τ and τ' are contained in τ^*), and there is a good deal of flexibility in choosing a logic L such that $S \subseteq \text{Sent}_L(\tau^*)$. Generally speaking, the extra sorts and relations in τ^* are necessary in order to connect together the domains and relations in models for τ with those in models for τ', since I have assumed that these two types are disjoint.

The above formulation of the argument is still imprecise because it is not yet specified what one means by 'the language' of a theory that is structuralistically described, i.e. presented in a language-free form. Taking the theory's potential models to be structures of a fixed similarity type, settles of course the matter of the theory's *vocabulary*. But there are as many 'languages' suitable for talking about structures of the given type as there are logics which admit that type. Thus, in a certain sense, to a theory T is associated not a single language, but rather a family of languages: to each logic L that admits τ one can consider for example the collection $\text{Sent}_L(\tau)$ of all L-sentences of type τ. Likewise, in order to say that a given statement φ is a *logical consequence* of T, in the sense of being true in each of its models $\mathfrak{M} \in M$, one must relativise to the logic L concerned; that is, indicate that $\varphi \in \text{Sent}_L(\tau)$ and that $\mathfrak{M} \vDash_L \varphi$ for all $\mathfrak{M}$ in M.

Though 'the language of T' cannot be identified with any single class of statements (indeed, to do so would break with the spirit of the structuralist view), evidently some languages do have a privileged status in this respect. This applies in particular to any language in which the theory's class of models M can be defined; that is to say, where for some L' and set $\Sigma \subseteq \text{Sent}_{L'}(\tau)$, $\mathfrak{M} \in M \Leftrightarrow \mathfrak{M} \vDash \Sigma$. In this case the 'laws' of T have an equivalent, linguistic formulation. Similarly, a special status should be afforded to a language $L(\tau^*)$ in which the reduction of T to T' is 'defined', and which thus yields a translation of $\text{Sent}_L(\tau)$ into $\text{Sent}_L(\tau')$. Obviously, the two logics L and L' just referred to need not be the same, though there is perhaps a particular interest in the case where they are or can be so taken.[16]

Picking up the threads of our argument once more, let ρ

reduce T to T', and let $\Gamma:\text{Sent}_L(\tau) \rightarrow \text{Sent}_L(\tau')$ be an induced translation that respects ρ in the manner of (5). Set $M'' = \text{Dom}(\rho) \cap M'$, and let T'' be a specialisation of T' that has M'' as its class of models. From (0) and (5) we get immediately

(9) $\forall\,\varphi \in \text{Sent}_L(\tau)\quad [\forall\,\mathfrak{M} \in M \quad (\mathfrak{M} \vDash \varphi)$
           $\Rightarrow \forall\,\mathfrak{M}' \in M''\quad (\mathfrak{M}' \vDash \Gamma(\varphi))].$

In the terminology just introduced, every L-consequence of T has a translation that is an L-consequence of T''. This general conclusion can be slightly strengthened in the case where the logic L also suffices for characterising the class of models of T (or where $L = L'$ above). Suppose for simplicity that M is defined by a single axiom $\theta \in \text{Sent}_L(\tau)$. Then, in virtue of

(10) $\forall\,\mathfrak{M}' \in M'',\quad \mathfrak{M}' \vDash \Gamma(\theta),$

we could say that a translation of (the axiom of) the reduced theory T logically follows from a specialisation of the reducing theory T'. If the class M'' is also axiomatised by an L-sentence, say $\theta'' \in \text{Sent}_L(\tau')$, (10) can be improved to

(11) $\theta'' \vDash \Gamma(\theta),$

so that T could be said to be deducible from T'' under translation. Again, 'deduction' here can be construed in the syntactical or proof-theoretic sense if the logic L is complete. Otherwise the logical relation of T to T' must be expressed under the broader rubric of semantical consequence, as given by (10) or (11). In any event, these considerations indicate that the structuralist framework, far from supplying a noninferential model of reduction, yields what essentially amounts to a generalisation of the standard deductive schema of reduction.

If correct, this last conclusion makes nonsense of Stegmüller's rationality thesis. At least as far as individual theories are concerned, structuralist reduction implies commensurability; for it

implicitly embodies just the kind of inferential link whose presence Kuhn, Feyerabend and others have denied in cases of
incommensurability.

What happens if one considers larger complexes of theories, or
even entire paradigms? Is there a chance to salvage Stegmüller's
thesis when it comes to scientific revolutions in the fullest sense
of the term? This situation is clearly more complicated to assess.
For one thing, Kuhn — at least in his earlier writings — has
maintained that paradigm incommensurability is a complex phenomenon arising from various *different* aspects of the disciplinary
matrix associated with a period of normal science. Thus, for
Kuhn, a new paradigm is accompanied, and partly constituted, by
fresh concepts, methods of research and norms for evaluating the
results of scientific inquiry. And incommensurability is accordingly understood to result from discrepancies in each of these
features of the paradigm. The absence of logical relations between
the theories or rival paradigms is therefore only one consequence,
or perhaps symptom, of incommensurability in Kuhn's original
sense.

It is unclear to what extent Stegmüller's reconstruction of
normal science and revolutionary progress succeeds in reproducing these broader aspects of paradigm change, hence whether
it might conceivably 'close the rationality gap' in a manner likely
to quell Kuhn's critics. One can readily see, however, that as far
as the rejection of *SR* is concerned, little alters if one switches
attention from individual theories to theory nets or evolutions;
in other words, if one turns to *paradigm* reductions in the
structuralist sense. In this context, the most important type of
theory net is called a *uniquely based σ-net*. Its distinguishing
feature is that it has a unique base theory, or maximal element
under the σ or specialisation ordering. Since all the theories in
such a net share the same class of potential models, one can associate to the net a single similarity type. A reduction of one net N

to another N' is then, according to Balzer and Sneed (1977/78), a reduction ρ of the base T of N to the base T' of N', such that every $T \in N$ is reduced by ρ to some $T' \in N'$. If the appropriate types are τ and τ', so that $\rho\colon \mathrm{Str}(\tau') \to \mathrm{Str}(\tau)$, then by our previous argument one should obtain, for some L, a translation $\Gamma\colon \mathrm{Sent}_L(\tau) \to \mathrm{Sent}_L(\tau')$ which respects ρ. One can thus regard N as being in this sense uniformly translatable into N'.

The situation is completely analogous in the case of paradigm changes. A period of Kuhnian normal science is represented by Stegmüller (1979) as a *theory evolution* $E = \langle \mathbf{N}, \leqslant \rangle$ comprising a collection $\mathbf{N}$ of uniquely-based σ-nets temporally ordered by $\leqslant$. The key feature of such evolutions is that the bases of all nets in $\mathbf{N}$ share a common *paradigm* core which thereby fixes the conceptual structure of E as a whole. A rival evolution E' may possess a paradigm core that is conceptually distinct from that of E. But when E' replaces E in a progressive and revolutionary change of paradigm, it is assumed in particular that the paradigm core of E is reducible to that of E', and that a distinguished net of E is likewise reducible to a corresponding net of E'. Again, one can infer that the reductions in question establish translatability relations between the relevant theories (or cores) of the two paradigms, and hence that incommensurability in Kuhn's sense cannot be a feature of this type of scientific change.

I have spoken so far about only one basic kind of reductive correspondence between theories, sometimes labelled *strict* or *exact* reduction. Structuralists do not intend this to exhaust all instances of theory replacement in science. They recognise in fact that an important class of theory-changes ought to be characterised under some model of what might be termed *approximative* reduction, where limits and approximations are involved in an essential way in relating the earlier theory to its successor. Though Stegmüller's 'defence' of Kuhn, and his central assertion *SR*, is constructed only on the basis of the strict reduction relation, he has in more recent writings emphasised the need to develop an adequate account of approximative intertheory explanation and reduction. And in this direction, the work of Moulines (1980) and Mayr (1981) provides the foundation for a definitive explication of this concept along structuralist lines.

A proper examination of this new notion of reduction would take us too far afield. However, the question does arise whether, by substituting 'reduction' with 'approximate reduction' in *SR*, a new and more adequate defence of Kuhn's theory might not emerge. I suspect that it will not, and I want to conclude this chapter with some remarks in support of this view. In the first place, an *approximate* reduction of a theory T to a theory T' is, according to Mayr (1981), an *exact* reduction of T to a so-called *completion* of T', denoted by $\hat{T}'$. $\hat{T}'$ is in a precisely defined sense an approximation of T', so one should at least be able to infer that the 'reduced' theory is commensurable with an approximation of the 'reducing' theory. Whether one can improve on this, and conclude that in such a case also T and T' are commensurable, depends to a large extent on how one views the approximation involved. If, for example, 'commensurability' is regarded as a transitive relation, and T' and its completion are deemed sufficiently 'close', one might argue, by transitivity, from the commensurability of T and $\hat{T}'$ to the commensurability of T and T'.

Such an argument would not, however, carry much weight for many supporters of the incommensurability thesis. For they would insist that, whilst in some cases an exact logical relation does hold between a theory and some approximation or idealisation of its successor (or, alternatively, between a theory and some approximation of its predecessor), one cannot conclude from this that the supplanting and supplanted theories are themselves commensurable. In Kuhn's analysis, for instance, the *apparent* commensurability of classical and relativistic mechanics is a myth disseminated by those who first modify and reformulate the classical theory in order to make it logically and empirically comparable with the superseding relativistic theory. For this reason an approximate reduction in the structuralist sense cannot *per se* count as a guarantee of commensurability; further argument is needed.

Though I can offer no watertight case for the claim that approximate reduction of this kind implies commensurability, strong evidence for it does emerge from the principal historical examples. The relation of Kepler's Laws to Newton's gravitation theory, and of classical particle mechanics to relativistic particle mechanics, have been reconstructed as instances of approximate reduction in the structuralist framework (see, e.g., Mayr, 1981; Stegmüller, 1986). For both these examples, which form the backbone of the structuralist explication, one can establish directly a relation of translation and deduction between the theories concerned. This can be achieved within a more general model-theoretic framework of the sort which Veikko Rantala and I have applied (cf. Pearce and Rantala, 1984a, 1984b, 1988); and in Chapter 7 I shall argue, on the basis of our characterisation of limiting case correspondence between theories, that in such cases of approximate reduction commensurability of the required kind is in fact forthcoming.

2

STRUCTURALIST CRITERIA
OF COMMENSURABILITY

My examination and rejection of Stegmüller's rationality thesis, encapsulated in his characterisation *SR* of scientific revolutions, led me into an analysis of various aspects of the structuralist view of theories in what might be called its *first* phase of development; that is, up to about the mid-1970s. Since that time, the structuralist programme has gathered remarkable momentum. On the one hand, it has generated many new case studies of theories and theory change in science. On the other hand, its underlying metascientific framework has undergone further enlargement and modification. In these more recent developments of the structuralist view, the problem of incommensurability too has come in for increasing attention. Whereas in the first phase incommensurability was treated as a more or less imprecise notion belonging to the old, statement view of science (as in *SR*), in its subsequent stages of evolution the structuralist approach has attempted to force even incommensurability into its set of rigorously explicated concepts. Early efforts in this direction, located in Balzer (1979) and Stegmüller (1979), were none too perspicuous; and our discussion will, I think, profit little by dwelling on them. But a fresh attack on the problem, in recent works by Wolgang Balzer (1985b), and by Stegmüller himself (1986), appears to be somewhat clearer and better motivated. The present chapter will be devoted to it.

I shall focus here especially on Balzer's (1985b), in which he sets out to achieve two basic goals: first, to refute the type of argument I gave in the previous chapter showing that a reduction relation in Sneed's sense can only hold between theories that are commensurable with one another; secondly, to set up within the structuralist framework precise criteria for the (in)commensurability of theories. These aims go hand in hand. The purported refutation of my argument is intended to lead to a natural and plausible account of incommensurability which, in turn, is taken to undermine the cogency of my argument.

As a whole, Balzer's analysis is also meant to yield additional support for *SR* as a characterisation of revolutionary scientific change. In this, it has received the full backing of Stegmüller, who has offered a detailed reconstruction and attempted rebuttal of my argument (in Ch. 10 of his 1986), which accepts an important part of Balzer's criticism. At the same time, Stegmüller provides a precise account of (in)commensurability which agrees in its main lines with Balzer's explication.

Though Balzer's and Stegmüller's discussions raise several important issues in the logic of theory change and reduction, they are, I believe, fundamentally misguided. I shall try to show here that neither of the main aims of their analyses is successfully achieved and that, accordingly, the evidence against Stegmüller's rationality thesis remains as strong as before. I shall begin with a brief resumé of Balzer's position on the commensurability question and his challenge to the argument of Chapter 1. I shall then attempt to rebut his challenge and defend my argument once again. Lastly, I shall consider in more detail the Balzer–Stegmüller explication of (in)commensurability, and show it to be in several respects unsatisfactory.[1]

1. BALZER ON INCOMMENSURABILITY

Balzer's reply to the argument of Chapter 1 and his new reconstruction of 'commensurability' are virtually inseparable, so they are best introduced as a single package. His strategy is to question the first and the third steps of my argument. He claims, first, that the assumptions (1.1)—(1.4) required in step one are too restrictive to apply to actual cases of reduction between scientific theories; so that the argument does not really get off the ground. And, secondly, he maintains that even if the 'initial conditions' could be made realistic, the argument would anyway balk at stage three, because the kind of translation guaranteed in step two is inadequate to establish commensurability; only a stronger form of syntactic connection would be sufficient for this.

In reconstructing my argument, Balzer tacitly accepts both assumptions (1.1) and (1.2). Actually, he strengthens the second of these by requiring that the reduction relation ρ is L-elementary, rather than L-projective, and by assuming that the logic L in question is adequate for both theories, T and T', in the sense that their model classes M and M' can be axiomatised by sentences of $L(\tau)$ and $L(\tau')$, respectively. As I already remarked, the first of these claims is unrealistic. The second is simply unnecessary, and may actually be damaging to the argument. Balzer's chief quarrel, however, is with premises (1.3) and (1.4), and with the final conclusion. He objects that one is not free to assume in a formal reconstruction that T and T' have disjoint types if in fact these theories, informally construed, happen to possess some terms in common. Further, he argues that, for the bulk of empirical theories employing ordinary mathematics, at least second-order logic will be required for an adequate formalisation. But then, by the well-known failure of Interpolation in second-order logic, it follows that L cannot be taken to comply with (1.4). Lastly, he holds that an adequate or *proper* translation between theories should preserve lexically similar terms as well as their references. In other words, a translation of the kind inferred in Chapter 1 will fail to establish commensurability if either $\tau \cap \tau' = \tau_0 = 0$, or if Γ is not an identity mapping on all sentences of $L(\tau_0)$.

From this last objection Balzer extracts general criteria under which a reduction ρ and an induced translation Γ respecting ρ would render T and T' commensurable. They are, in slightly simplified form, as follows:

(0)   (i) $\mathrm{Symb}(\tau_0) \neq 0$;
        (ii) $\forall\, \varphi \in \mathrm{Sent}(\tau_0),\quad \Gamma(\varphi) = \varphi$;
        (iii) $\forall\, R \in \mathrm{Symb}(\tau_0),\quad \exists\, \mathfrak{M}' \in \mathrm{Dom}(\rho),$
              $\mathfrak{M}'(R) = \rho\,\mathfrak{M}'(R).$

The first condition requires that the common subtype τ_0 of τ and τ' has a nonempty vocabularly (assumed for simplicity to consist of relational symbols). The second says that Γ can be chosen to be an identity mapping on the shared sentences; in other words Γ is a homophonic or, what I prefer to call, *literal* translation.[2] The third condition demands that for every predicate R common to both theories, there is at least one pair of models $(\mathfrak{M}', \mathfrak{M}) \in \rho$, such that R has the same extension in both $\mathfrak{M}'$ and $\mathfrak{M}$. (The pair in question might be different, for different predicates R.)

Balzer's notion of commensurability thus includes a semantic and syntactic component. For T and T' to be commensurable there must first of all be a reduction ρ of T to T' (or conversely) fulfilling Sneed's criterion (1.0), together with a full translation Γ satisfying (1.5) above. The translation must be literal or homophonic, in the nonvacuous sense that there are some sentences shared by both languages. And the model-theoretic correspondence should establish some referential overlap between the shared terms, as laid down in (0)(iii). Otherwise, if the theories have disjoint types, or if no suitable ρ and Γ exist, they are held to be *incommensurable*. This applies even if one of the theories is reducible to the other in the structuralist sense. It also provides a challenge to the last step of my argument, since there is no guarantee that the translation whose presence I inferred will generally satisfy the additional conditions (0)(ii) and (iii).

2. A RESPONSE

Is Balzer's critique of my argument in any sense sound? Does the ensuing characterisation of (in)commensurability achieve any genuine clarification of this concept? The correct answer on both counts is, I am convinced, negative. But to come to a measured assessment of Balzer's position one needs to consider a number of related but wider issues in the logic of theory reconstruction and scientific change.

The first striking feature of Balzer's proposals is that he takes matters of language and translation to be central concerns in the enterprise of understanding (and precisely reconstructing) the phenomenon of incommensurability. This, on any account, is uncontestable. Indeed, perhaps the major thrust of my argument in the last chapter was to underscore exactly this point: that the structuralist approach is inadequate when it comes to analysing questions of comparability and commensurability when these are basically conceptual and linguistic problems. And, I claim, it is largely due to this weakness of the 'nonstatement view' that Stegmüller is led astray in his 'defence' of Kuhn. Balzer's reversal of the usual structuralist stance here is no mere minor-concession-to-the-opposition, for-the-sake-of-the-argument strategy. It amounts to a quite drastic paradigm-shift on his part. And if his proposals become accepted structuralist lore, one will no longer be able to speak as before of a peculiarly 'nonstatement' perspective at all, nor insist that the structuralist view can provide a radically alternative model of scientific growth.

Having made the switch to the opposition's 'paradigm', Balzer then tries to bring linguistic considerations to bear on the problem of theory comparison, at both syntactic and semantic levels. The chief motivation behind his characterisation (0) seems to be this: that theories are incommensurable if they employ some words with sufficiently different 'meaning'. In the context of a Sneed reduction of T to T', this difference of meaning is explained in clause (iii) of (0) by saying that a given, shared term R has nowhere in the two theories the same reference; or that R denotes a different relation in all models of the theories that are connected by the reduction function. Syntax then comes into the picture because a commensurability-inducing (or *proper*) translation is supposed to preserve identical words, besides allowing for some overlap of their reference.

Later, I shall argue that Balzer's constraints on denotations and 'proper' translations are unacceptable. For the moment, however, let us consider his critique of my argument in more detail.

It begins with the observation that generally the two theories T and T' under comparison will possess some terms in common. Though this point is not very fully developed, it is taken to be a serious objection to my assumption (1.3) of disjoint types. Suppose, however, that the relevant types τ, τ' have some nonempty, common subtype τ_0, and that one has available as before a reduction ρ of T to T'. There are various ways to proceed. One might entertain the (admittedly rather strong) assumption that all terms in τ_0 have exactly the same meaning in T as in T' and that this meaning-invariance is respected by the model-theoretic operation ρ. One could not apply Feferman's uniform reduction theorem as it stands; but if one considers the identity mapping on τ_0-sentences as fixing part of the translation in advance, one could then try to combine suitable strengthenings of the requirements on ρ, together with the remaining assumptions (1.1), (1.2) and (1.4), in order to infer the presence of a translation which acts on all sentences. As a step in this direction, one can show, for example, that the consequent of Feferman's theorem remains valid when, in its antecedent, the condition $\tau \cap \tau' = 0$ is replaced by the weaker condition that, for every sort j common to τ and τ', if $(\mathfrak{M}', \mathfrak{M}) \in \rho$, then $\mathfrak{M}'(j) = \mathfrak{M}(j)$.[3]

Another way to proceed is to apply my original argument by renaming one of the types; say, by starring all symbols and sorts in τ. This would give us a uniquely determined theory T^*, say, which is merely an alphabetical variant of T. The only conceivable objection that might arise would be if one could prove some interesting property of T^*, for instance its commensurability with T', that one could not demonstrate for T. But one cannot. For, let ρ be the given reduction of T to T', let ρ^* be the induced operation from $\mathrm{Str}(\tau')$ to $\mathrm{Str}(\tau^*)$, and suppose that assumptions (1.1)–(1.4) hold with respect to ρ^*, T^* and T'. Obviously, ρ^* reduces T^* to T' and, by the argument, yields (for some L) a translation Γ^* of $\mathrm{Sent}_L(\tau^*)$ into $\mathrm{Sent}_L(\tau')$. Now, from the assumption that L admits name-changes it follows at once that, since Γ^* respects ρ^*, one can define a translation $\Gamma: \mathrm{Sent}_L(\tau) \to \mathrm{Sent}_L(\tau')$ which respects ρ: just set $\Gamma(\varphi) = \Gamma^*(\varphi^*)$, for all φ, where φ^* is the notational variant of φ. It is then easily verified that Γ and ρ satisfy (1.5).[4] In other words, for any translation that is obtained in this manner by renaming, there is a translation between the theories in their original form. Contrary to Balzer's contention, in the present context the relabelling of primitives is a harmless pastime.

His second objection is less straightforward to dispense with. It is directed at assumption (1.4), that the logic L which projectively defines the reduction can be thought to possess the Interpolation Property. His counter-argument appeals to what Balzer calls the *empirical* philosophy of science; that is to say, the domain of real scientific theories to be reconstructed as actual or potential cases of reduction. Real empirical theories generally embody strong mathematical systems, like analysis and arithmetic, containing second-order axioms: completeness, induction, and so forth. An adequate logical reconstruction ought to bring out the full strength of the underlying mathematics, and will therefore require the use of higher-order logics for which the Interpolation theorem is known to fail. Consequently, he argues, for the majority of concrete cases of reduction in science, assumption (1.4) will be invalid.

The above claim is developed from two sides. Balzer first requires that the full strength of the underlying mathematical systems be exhibited in the reconstruction. His second demand is that *only* the mathematics explicitly present in an empirical theory should be formalised; this is intended to block the possibility of appealing to set theory in order to obtain a first-order formulation of the axioms. For, he insists, "only those axioms should be made explicit which are treated explicitly in the informal theory itself". This yields a partial criterion of adequacy for logical reconstructions which one could express as follows:

> *The reconstruction must reproduce the actual mathematical axioms employed in the empirical theory.*

But this criterion is, to my mind, quite unworkable and would, if generally applied, lead to unnecessary and undesirable restrictions on the ways in which scientific theories and intertheoretic relations are rationally reconstructed.

Before turning to a deeper examination of this criterion, however, there is a general point to be made about Balzer's objection to assumption (1.4). As noted earlier, he seems to assume that the logic L 'defining' the reduction is also sufficient for axiomatising the theories T and T' concerned; the presumption is that M and M' are L-elementary classes of models. I pointed out in the last chapter that this premise is neither required in order to infer commensurability on the basis of structuralist reduction, nor is it in general a defensible hypothesis. By insisting on it, Balzer is in effect demanding a constraint on our formalisations that his own approach willingly forswears. It has been a recurrent theme of the structuralist literature that the set-theoretic mode of reconstruction is able to by-pass certain logical and linguistic features of empirical theories to gain a more 'direct' access to intertheory relations. On their account, a reduction relation between, say, two *physical* theories will have a particular 'physical' significance whose content can be represented without probing into technical details of the underlying logical and mathematical bases of the theories. If correct, this should entail that when logically reconstructing a given Sneed reduction, what should concern us is how faithfully the 'physical content' of the relation is reproduced. It is perfectly feasible that one obtains a full linguistic characterisation of the *relation* without providing a comprehensive *formalisation* of each *theory* (in the syntactical sense). In principle, at least, the content of a reduction might be adequately expressed within a first-order language (or another language possessing Interpolation) *independent* of whether the theories involved admit an elementary formalisation. Balzer's argument is thus directed against the wrong target, for he attempts to show that empirical *theories* must be necessarily second-order theories, instead of showing that *relations between them* must be second-order relations.

Even if Balzer's argument were not misdirected, it would, I believe, lack any real bite. The first problem with his fidelity criterion is that it assumes scientific theories to possess a more or less *explicit* mathematical basis. The assumption is less innocent than it appears. Let us for the moment set aside the 'extreme' cases, where theories require only minimal mathematics, or where they employ some strong set-theoretic formulation. Let us grant that typically real and complex analysis in some form or another underlies quantitative, mature empirical theories. Even so, the fidelity criterion would be hopelessly hard to apply. Most presentations of a scientific theory include no explicit axiomatisation of the mathematical base, and many leave the mathematics entirely implicit. Naturally, chapters and appendices devoted to the elements of set theory and analysis often intrude into textbook treatments of science (they are usually absent from research papers). This is an aid to the reader in becoming familiar with mathematical concepts and methods used in the text. But there is rarely any concerted effort on the part of the author to trace the specific dependencies of the problem solutions on the stated mathematical axioms or assumptions, much less to develop a feeling for when a given mathematical presupposition might be relaxed or weakened. The chief concern of textbook writers is didactic clarity, which usually means they are content to adopt as much heavyweight mathematics as will yield straightforward and routine solutions to empirical problems. Questions of logical finesse, and how to reduce mathematical assumptions to a minimum, thus hardly arise.

It seems clear, in any case, that the criterion of fidelity cannot be applied to an 'informal theory' in the usual sense of the term. It could at best be applied to specific *formulations* of a theory, leaving us with the problem: When are two formulations versions of the same theory? The natural response: "when they have equivalent logical reconstructions", would be plainly question-begging in this context. Moreover, if taken at face value the criterion would encourage one to select as preferable those reconstructions which differ as little as possible from the original theory, thus defeating the whole enterprise of reconstruction. The purpose of a reconstruction is, after all, to fill gaps in an original, to flesh out logical structure, supply missing links and repair logical and conceptual deficiencies. This will be carried out against a background of specific logical, methodological and epistemic assumptions and goals. The aim, as well as the ultimate success, of a reconstruction is therefore highly context relative. It is, accordingly, a *selective* process: just as it sorts, shuffles and reorganises existing features of a theory, incorporates new ones, and makes explicit hidden assumptions, so it may suppress or eliminate features that are considered irrelevant to the aims at hand. This is unavoidable if the reconstruction is to function as a workable 'technical unit'.

The fundamental weakness of Balzer's criterion is perhaps best illustrated by means of a simple example. Consider any physical theory that makes essential use of the real number system. Looked at from a completely 'informal' or 'naive' perspective, one could well imagine that the 'real' reals are intended to form the mathematical base of the theory. Yet as soon as the theory is axiomatised in any extensional logic, models in which isomorphic copies of the reals appear will be admitted. This is scarcely a cause for alarm, because any isomorphic models generated by one's reconstruction are logically equivalent to one another, and their presence has no bearing on the cognitive content of the theory. Categorising the models *up to isomorphism* is thus the small price one pays for being able to define them in a well-behaved logic. All the same, there is a sense here in which one *could* claim infidelity to the 'original' theory, and hence conclude that virtually any *logical* reconstruction will have to break with the spirit of Balzer's criterion.

I prefer a different interpretation, however. It seems to me that what this trivial example shows is not that the criterion is 'false' or too strict, but that in this context it cannot be made sense of. In fact, what should be questioned here is the initial presupposition that the 'real' reals are somehow intended by the theory. For, the *physical theory*, as I see it, simply does not make a pronouncement on this issue, one way or the other. It is our logical reconstruction, rather, that resolves a feature that is left vague in the original theory. Once again, one need only consult textbooks and journals of physics to see that the levels of logical sophistication and mathematical formalisation are generally quite low. Not only is the mathematics often imprecisely stated, from time to time one finds mathematical passages and inferential steps that are, by normal standards of rigour, illegitimate.

To sum up, the adequacy of a logical reconstruction should be assessed relative to its specific goals. Absolute fidelity (when this can be made sense of) is seldom a contributive factor in the assessment, and even approximate fidelity is rarely decisive. One might, however, consider other kinds of criteria: for instance, the criterion of *sufficiency*, according to which what matters is not how much mathematics is actually assumed by some or all formulations of an empirical theory, but how much *needs* to be assumed for the theory to perform its usual cognitive function. This requirement, again, is vague, but when context is taken into consideration it can sometimes be cast in a reasonably sharp form. In the case of reduction, for example, one might say that an adequate reconstruction ought to reproduce as many of the logical and mathematical presuppositions as required to define and illuminate a given, informally presented reduction relation. Arguments for or against a particular choice of logic in this context would then turn on certain theoretical criteria of sufficiency, besides pragmatic criteria of clarity, uniformity, and so on.

I have already remarked that logical reconstructions are not as a rule required to provide a *rigid* characterisation of the reals. Should they give a categorical characterisation? In other words, must the reals be defined, up to isomorphism, by a *standard* model of analysis? One possible argument for a categorical representation of the reals derives from considerations of measurement. The ordinary real number system offers a simple and powerful framework for quantitative theories because it permits physical magnitudes to be represented on a continuous scale and allows one to characterise relations between physical quantities, and indeed physical laws generally, by means of functions that are continuous, differentiable, and so forth. One argument here in favour of the standard real number system stems from empirical measurement. One can lay down exact postulates governing fundamental, qualitative measurement, and then show that the so-called *extensive* measurement structures which satisfy the postulates can be embedded into numerical relational structures, and hence into models of analysis. Algebraic representation and uniqueness theorems establish a strong connection between the basic measurement models and the standard model of the reals, endowing the latter with a sort of 'empirical' foundation.[5]

However, it is well known that the only postulate of fundamental measurement that cannot be expressed by a first-order sentence — viz. the Archimedean axiom — has a different status from the remainder. Unlike the other axioms, it cannot be empirically grounded because in a precise sense it cannot be refuted by any finite amount of observational data.[6] It seems, therefore, that its 'justification' has to rest on pragmatic grounds, or else on the basis of its conformity with ordinary usage. But in fact the cost to measurement theory of omitting the Archimedean axiom may well be minimal; for it can be shown that analogous representation and uniqueness theorems reveal an equally strong connection between models of the remaining postulates and certain *nonstandard* models of analysis that contain nonstandard, elementary extensions of the reals.[7] This shows that from an empirical point of view the nonstandard models are equally 'well grounded'. If this is coupled with the fact that, technically, nonstandard analysis is a rich and fruitful mathematical theory with direct application in the special sciences, in ancillary disciplines like probability theory, and in metascience generally, it seems to follow that a standard, categorical representation of the reals is neither empirically necessary nor, in all cases, methodologically desirable. And this of course undermines the motivation for insisting on second-order axiomatisations.

The argument just given can I think be adapted to counter the view that also the *geometry* of physical theories must be categorically characterised by a fixed structure, such as that of

Euclidean space: $\mathbb{R}^4$, say. It is well known that the axioms of elementary geometry do not determine this space up to isomorphism. But one should also inquire whether this customary representation of Euclidean space is an intrinsic and necessary feature of classical and special relativistic physics, or whether its adoption is not rather a matter of habit and convention. I do not want to discuss this question in detail here. But it is worth pointing out that lately several writers have argued against the direct identification of space—time with $\mathbb{R}^4$, and in favour of constructing space—time out of some primitive notions from affine geometry which can be empirically interpreted as relations between physical objects and events. What these writers urge is not so much a reform of theory as a novel method of theory reconstruction. The primary advantages are that space—time theories may receive a more direct physical interpretation, that relational theories of space and time can be more accurately depicted, and that different space—time theories can be more perspicuously compared.[8] As in the case of fundamental measurement, representation theorems are employed to show that the basic strength of ordinary space—time structures is reproduced by models in which only 'empirical' relations between 'physical' objects occur.[9]

At the present time it is still an open question which logics are most appropriate for characterising such 'empirically interpreted' geometric structures. But two points emerge quite clearly. First, it is by no means obvious that higher-order logics will be needed to define the requisite space—time theories. And, secondly, it is already apparent that the work of Field and others in this area has achieved considerable progress towards reducing the mathematical ontology associated with physical theories. Moreover, this kind of work is plainly relevant to the study of intertheory relations in general. Far from welcoming these developments, Balzer's criterion effectively renders them idle as contributions to the "empirical philosophy of science".

There is also a stronger and more compelling rejoinder to Balzer's argument. Even if one were to insist, as he does, that in physical theories only a categorical characterisation of the reals is legitimate, it does not follow that only second-order logic or a logic without the Interpolation property would suffice for this. The standard reals can be defined not only in second-order logic, but also in infinitary logics $L_{\kappa\kappa}$ ($\kappa > 2^{\omega}$) which do not possess higher-order quantifiers. (In fact, the completeness of the reals can be defined in an uncountable first-order language that has a name for each real number.) Generally speaking, Interpolation fails in large infinitary languages, but recent developments in infinitary logic have shown that this is not always the case. A result of Oikkonen (1985) shows that Interpolation does hold for some of the so-called *infinitely deep* (or Hintikka—Rantala) languages, $M_{\kappa\kappa}$. These logics are 'infinitary' in a different sense from the usual one; however, in general $M_{\kappa\kappa}$ has a greater expressive power than $L_{\kappa\kappa}$. It follows that the reals can be characterised in logics with Interpolation, and hence that Feferman's theorem can be applied to infer translatability even when Balzer's constraints are met.[10]

3. ADEQUACY OF TRANSLATION AND MORE
ON UNIFORM REDUCTION

At this juncture I want to interrupt briefly discussion of the acceptability of assumptions (1.1)—(1.4) in order to bring into the picture Balzer's second main objection to the argument for commensurability. His quarrel, you may recall, is with the third step: the inference from translatability to commensurability. And his claim, in short, is that even if one *could* deduce the presence of a translation of sentences, Γ, which respects the model-theoretic correlation ρ, Γ might nevertheless fail to be, in his words, a *proper* or adequate translation.

This objection is hard to make sense of, since it seems to fly in the face of any standard semantic theory of translation. Later, in Chapters 6 and 7, I shall take a closer look at properties of translations and at possible desiderata which adequate translations should satisfy. For the moment, however, the following point is worth stressing. According to any reasonable semantic approach, a translation from one language to another is a mapping of expressions (of given semantic categories) to expressions (of the same or corresponding categories) which preserves *semantic values*. What kinds of values these are depends on the types of languages involved and the way their interpretations are specified. For suitably formalised empirical theories, for example, the values would normally be given by the (set-theoretic) extensions of closed terms and formulas. But, whatever the relevant semantic values happen to be, the adequacy of a translation will always be assessed *relative* to some correlation between the models (or interpretations) of the two languages; otherwise, one could not say with respect to *which* interpretations of the languages the appropriate values had been preserved. Hans Kamp sums up the standard conception as follows:

> A *translation* from a language L into a language L' is defined as a certain function from the syntactic analyses of expressions (in particular, sentences) of L to syntactic analyses of expressions of L' of the same semantic type; the *adequacy* of a translation is defined relative to a certain fixed association of a model for L' with each of the models of L. Relative to such an association a given translation T is adequate if:
>
> (1) it transforms each L-expression φ in its domain into an L'-expression $T(\varphi)$ such that for each model $\mathfrak{A}$ for L, $\mathfrak{A}$ assigns to φ the same semantic value as the model for L' associated with $\mathfrak{A}$ assigns to $T(\varphi)$; and
>
> (2) the domain of T includes all (analyses of) sentences of L.
>
> (Kamp, 1978, p. 278)

This very general formulation is not intended to provide a

definition of translations for particular languages, but rather to give a broad characterisation under which the concept of adequate translation is to be subsumed in concrete cases (rather in the manner that Tarski's 'Convention T' lays down constraints on truth definitions). Kamp's own special interest is in the cases where L and L' are *intensional logics* in Montague's sense, or where they are formalised fragments of natural languages. But his characterisation is readily applicable to the situation of reduction between scientific theories: for 'L' and 'L'' above, read 'L(τ)' and 'L(τ')' respectively, and let ρ be the usual mapping which associates models for $L(\tau')$ and $L(\tau)$. Now, suppose that the satisfaction relation $\models$ for L is inductively defined on closed terms and thence formulas, and that the syntactic analyses available in translation go down to the level of *terms* of $L(\tau)$. For any term t and model $\mathfrak{M}$, denote by $[t]^{\mathfrak{M}}$ the interpretation (extension) of t in $\mathfrak{M}$. As before, let Γ be the translation function. Then Kamp's adequacy condition (1) might be expressed by saying that for every closed term t of $L(\tau)$ and model $\mathfrak{M}'$ in the domain of ρ,

$$[\Gamma(t)]^{\mathfrak{M}'} = [t]^{\rho(\mathfrak{M}')}.[11]$$

In the more general setting of Chapter 1, where sentences are the basic lexical units and the semantic values are *truth values*, this condition might be reformulated by

$$\mathfrak{M}' \models \Gamma(\varphi) \Leftrightarrow \rho(\mathfrak{M}') \models \varphi,$$

for each $\varphi \in \mathrm{Sent}_L(\tau)$; and this is just a repetition of the requirement (1.5) that Γ respect ρ.

As far as my argument is concerned, just *how* the model-theoretic correlation ρ is defined is of marginal interest, because the adequacy or propriety of Γ is assessed *relative* to ρ. Thus, for present purposes it makes little difference whether ρ is determined as in Balzer (1985b) by a reduction relation, or, as in Balzer (1985a), on the basis of an association of $\mathrm{Str}(\tau)$ and

Str(τ') with a set of "real (empirical) systems" (eine "Menge realer Systeme") that these classes of structures are supposed to 'model'. In either case, if ρ is an acceptable correlation, then Γ is adequate to the extent that it respects ρ. This is all that the semantic theory of translation demands. Conversely, if the adequacy of a respect-ful translation is to be questioned, it can only be on some other, nonsemantic grounds, or else because ρ itself is unacceptable. The second alternative is not available to someone like Balzer who wants to defend the adequacy of a structuralist correspon-dence ρ. The first alternative might appeal to those wishing to invoke additional, broader (e.g. pragmatic) criteria of commen-surability. But Balzer does not take this step. His extra constraint on translation is essentially *syntactical* ((0)(ii)), and it is, as I shall argue below, quite unsatisfactory.

There is, however, another line of criticism that might be developed against the adequacy of our translation. It is often suggested in connection with formal (rather than natural) lan-guages that translation should be in some sense an *effective* procedure. Should it not be the case, therefore, that, given any arbitrary inputs of $L(\tau)$-formulas, one ought to be able effectively to compute their respective $L(\tau')$-translations? And if, as I conceded in Chapter 1, effectiveness is not ensured, is that not ground enough to query the entire argument from reducibility to commensurability? This conclusion would, I think, be far too harsh.[12]

To see why, let me take the opportunity to reconsider Balzer's distinction between *empirical philosophy of science* and what one might term *general metascience*. To illustrate the difference in the present context, one might say that whereas general metascience has the task to formulate, classify and study the properties of various general types of intertheory relations (including reduc-tion), the empirical philosophy of science is concerned with discovering and exploring the relations that actually hold between particular scientific theories. Obviously, there is, or should be, a close interaction between the two disciplines: the former seeks to develop concepts, methods and general explications on the basis of which actual case studies from science can be precisely analysed and reconstructed. Historical case studies, in turn, yield raw materials for motivating and 'testing' the general explications.

Though Balzer's distinction is perhaps more indicative of a difference in philosophical emphasis, rather than doctrine, it can provide a useful grip on understanding the approaches of different authors. One can see, for example, that Suppes' work is chiefly devoted to empirical philosophy of science, whereas Sneed's belongs more to the realm of general metascience. This, I think, explains their slightly different approaches on issues like reduction. For both writers, reduction is basically characterised as a structural or model-theoretic link ρ; but Suppes' assumption that in each case $\rho(\mathfrak{M})$ will be *constructible* (or, as I ventured to add, *definable*) from $\mathfrak{M}$ has vanished from the later accounts of Adams and Sneed.[13] Suppes' view can easily be defended at the 'empirical' level by observing that when reduction relations are presented for specific historical examples, one obtains or expects to obtain a concrete construction of $\rho(\mathfrak{M})$ from $\mathfrak{M}$. Sneed's account is potentially the more general of the two. But, unless one has a specific notion of 'construction' in mind, one cannot really infer that Sneed would actually admit reductions that Suppes excludes.

The lesson to be drawn from this can be spelled out quite simply. Agreed, Feferman's type of uniform reduction theorem does yield a very loose connection between sentences. This is the price one pays for exploiting a highly general result and inserting the weakest possible information about the ρ-relation, namely that which is minimally required to make it comply with Sneed's explication. I added only the plausible premise that ρ is 'definable' in a weak sense. But in genuine instances of reduction in science, one can certainly expect the ρ-relation to be determined by some sort of explicit construction of T-models from T'-models, and therefore that the induced translation Γ yields some sort of recursive or effective mapping of sentences. In other words, precisely at the level of empirical philosophy of science one has firm grounds to believe, contrary to Balzer, that a Sneed reduction establishes an explicit and adequate translation.

As I remarked in the last chapter, there are various different sets of semantic assumptions about ρ that can be invoked to infer translatability via the medium of uniform reduction. Whilst to my knowledge all these sets of assumptions are, on balance, stricter than the conjunction of (1.1) to (1.4), each of them provides more explicit syntactic information about the nature of the induced translation. Moreover, they can be chosen such that some of our individual assumptions — like (1.4), for instance, or even (nominally) (1.2) — are relaxed. What follows is a limited sample. In each case, I retain the notation and terminology of Chapter 1 and assume that ρ is a Sneed reduction of T to T', that (1.1) and (1.3) hold, and, where appropriate, that S is a set of τ^*-sentences projectively defining ρ, as in assumption (1.2).

(A) Suppose that ρ determines a *Gaifman operation* $G: \text{Str}(\tau') \rightarrow \text{Str}(\tau^*)$ in the sense that (i) $\text{Dom}(G) = \text{Dom}(\rho)$, and (ii) $\forall \mathfrak{M}' \in \text{Dom}(G)$, $(\mathfrak{M}', \mathfrak{M}) \in \rho \Leftrightarrow \exists \mathbf{R}, \langle \mathfrak{M}', \mathfrak{M}, \mathbf{R} \rangle = G(\mathfrak{M}')$. In particular, $\tau^* = \tau' \cup \tau \cup \{\mathbf{R}\}$ (where $\mathbf{R}$ is a collection of relations), S is a theory in $L(\tau^*)$ which defines the range of G, and there is a theory S' in $L(\tau')$ defining the domain of G (i.e. $\text{Dom}(\rho)$ is L-elementary). G is a single-valued algebraic operation in the sense of Chapter 1, and it can be represented by the pair (S', S). If, further, G is a *rigid* operation (meaning that $G(\mathfrak{M}')$ has no nontrivial automorphisms over $\mathfrak{M}'$) and L is first-order logic or a countable admissible logic, then, by a theorem of Gaifman (1974), there is a *defining schema D* such that, for every $\mathfrak{M}'$, $G(\mathfrak{M}')$ is *defined in $\mathfrak{M}'$ by D*. Gaifman's defining schemas generalise the usual notion of explicit or Beth-definability (and weaker versions thereof) to the situation where new objects as well as new relations are 'defined'.[14] His result generalises Beth's theorem by showing that a certain semantic or implicit definability of this kind entails syntactic definability too. And the new theorem implies the existence of a recursive translation from $L(\tau^*)$ into $L(\tau')$. (A similar result has been credited to Shelah for $L = L_{\omega_1\omega}$.)[15]

(B) Let ρ be a *$\varkappa$-word-construction* from τ' to τ (for $\varkappa$ a regular cardinal). The precise meaning of this notion is too involved to be repeated here; and for details the reader is referred to Hodges (1975). However, word-constructions are in a certain sense special cases of Gaifman operations, though they embody a generalisation of Gaifman's defining schemas to infinitary logics. It follows from Hodges' uniform reduction theorem (Theorem 11 of Hodges, 1975) that for every sentence φ of $L_{\infty\varkappa}(\tau)$ there is a sentence $\Gamma(\varphi)$ of $L_{\infty\varkappa}(\tau')$ such that $\forall\mathfrak{M}' \in \mathrm{Str}(\tau')$, $\rho(\mathfrak{M}') \vDash \varphi$ iff $\mathfrak{M}' \vDash \Gamma(\varphi)$. Theorem 13 of Hodges (1975) indicates that the translation Γ here is, in a suitable sense, *recursive*. Notice that we have now dispensed with assumption (1.4), because in general there is no interpolation theorem for the logics $L_{\infty\varkappa}$.

(C) Suppose that $\mathrm{Dom}(\rho)$ is an $L_{\omega\omega}$-elementary class (i.e. definable by a first-order theory) and that ρ preserves Fraissé partial isomorphisms in the sense of Sette and Szczerba (1978). It follows from a result of Sette and Szczerba that there is a recursive translation of $L_{\omega\omega}(\tau)$ into $L_{\omega\omega}(\tau')$ which respects ρ. Moreover, if T and T' are first-order theories and $\mathrm{Dom}(\rho)$ is exactly the class of models of T', then this translation determines an elementary interpretation of T in T'.[16]

(D) Similar to (C) except that ρ defines a *reduction functor* from M' into M in the sense of van Benthem and Pearce (1984), where ρ preserves isomorphisms and ultraproducts. Again, by their result, it follows that the first-order theory of M is (generalised) interpretable in the first-order theory of M'. In this case, as in (C), the assumption (1.2) that ρ is a definable relation, is replaced by the condition that the domain of ρ is first-order definable and that this mapping satisfies some standard mathematical (preservation) properties; the latter then imply the definability of ρ, and of T-models in T'-models.

The assumptions required in these last two cases may seem unduly restrictive in that they apply to first-order theories or to reduction relations possessing an essentially first-order 'content'. This may limit their sphere of influence to some extent. On the other hand, the results referred to demonstrate that the familiar syntactic relation of interpretability between elementary theories, for so long a cornerstone of the classical approaches to reduction in empirical science as well as in mathematics, can be given what virtually amounts to an *algebraic* characterisation. This makes them highly relevant to the structuralist programme which seeks to express the substantive cognitive content of intertheory relations in exclusively structural or mathematical form. It means, in effect, that this aim *is*, to a limited degree, realisable, *if* one wishes to retain something like the standard intuition of reduction. If, on the other hand, one is hoping (as Stegmüller appears to believe) that the structuralist framework is going to yield a radically alternative model of reduction, then the results cited above are a somewhat sobering reminder that — if rather simple kinds of mathematical conditions on structures are fulfilled — a Sneed reduction induces a very strong inferential link between theories, after all.

Gaifman's single-valued operations and Hodges' word-constructions have the rather different scope of providing model-theoretic characterisations of some standard types of operations and constructions performed in algebra. Their results yield explicit syntactic information about how properties of the constructed objects (algebras, topological spaces, and so forth) are reduced to or determined by properties of the structures that one operates on. They are not restricted to constructions expressible in first-order logic, nor is their domain of application confined to mathematical theories. Any example of reduction in science that involves performing an algebraic type of construction to obtain models of the reduced theory from models of the reducing theory is fair game for this approach.[17]

The concepts and methods developed by Feferman, Gaifman, Hodges and others contribute therefore to the problem I mentioned earlier in connection with Suppes' account of reduction: the problem of explicating what is to count as a *construction* in reduction contexts. And their results indicate different ways in which translatability phenomena arise in intertheory relations. There is no need to insist that these methods provide an exhaustive treatment of the subject; or to suppose that all instances of reduction fall under one of the sets of assumptions here listed. In fact, I shall discuss much later on a type of model-theoretic operation, also accompanied by syntactic translation, that does not *prima facie* comply with the stated assumptions: an operation which forms standard approximations of nonstandard models of particle mechanics.

To recap, the force of the argument of Chapter 1 was to show that *when* certain, rather general and plausible semantic constraints on reduction are met, translatability is assured. We have seen now that when these constraints are strengthened in different ways, the result will be correspondingly stronger forms of translation. To weaken or alter these constraints in other respects may mean to forgo the kinds of uniform reduction theorems that yield translatability as an *automatic* consequence. But it certainly does not imply that untranslatability will be a necessary or even likely outcome. I suspect that *all* model-theoretic operations and constructions that Sneedians are prepared to regard as establishing valid reductions will as a matter of fact be accompanied by an adequate intertheoretic translation. And if this conjecture turns out to be mistaken, I think one will have to examine the cases in question very carefully indeed and ask whether something is not sorely amiss with the structuralist construal of reduction.

4. THE STRUCTURALIST CRITERIA REJECTED

I want to turn now to Balzer's own characterisation of commensurability, as given in (0) above. As one might have expected, this explication reverses some of the standard intuitions about reduction and commensurability. Normally one thinks of reduction as a relation holding only between commensurable theories, whereas now commensurability is interpreted as a relation holding only between pairs of theories, one of which is reducible to the other. In line with his defence of Stegmüller's rationality thesis, Balzer regards some reductions as establishing commensurability, others not. The difference is supposed to be due to the type of translation involved.

As we saw, Balzer's chief adequacy criteria for translations to establish commensurability are, first, that they respect an antecedently available Sneed reduction, and, second, that they are *literal* (condition (0)(ii)). Actually, his requirement is slightly stronger than this: Suppose that T is reducible to T' by a correspondence ρ, that R is any relation symbol common to the vocabularies of T and T' ($R \in \tau_0$), and that Γ is a translation of Sent(τ) into Sent(τ') which respects ρ. Then, in his terminology, Γ is *R-preserving* if for any sentence φ in the domain of Γ, all occurrences of R in φ remain unchanged after translation. (If all shared vocabulary is preserved in this sense, then clearly the translation is literal, and (0)(ii) holds.) Further, Γ is said to *render*

T and T′ R-ρ-commensurable if the equality expressed in (0)(iii) holds — that is, if for some $\mathfrak{M}' \in \mathrm{Dom}(\rho)$, $\mathfrak{M}'(R) = \rho\mathfrak{M}'(R)$. *T* and *T′* are then held to be *commensurable* if there exist suitable ρ and Γ, such that for *all* shared symbols *R*, Γ is *R*-preserving and renders the theories *R-ρ*-commensurable, i.e. (0) is satisfied (modulo the slight difference mentioned above). The theories are of course *incommensurable* otherwise.

This construal of (in)commensurability is puzzling in several respects, and has a number of counter-intuitive and unacceptable consequences. One serious drawback, as I see it, is that an existing reduction relation is taken as a *precondition* for theories to be commensurable. It is not only undesirable to restrict the domain of commensurable theories in this way, it is also quite unnecessary for the purpose of defending *SR* by showing that at least some Sneed reductions obtain between incommensurable theories. However, leaving this objection aside for the present, there is, I think, an extremely simple argument to show that Balzer's chief constraint on translation, namely that of literality, cannot elucidate the concept of commensurability in any reasonable sense of the term.

Whatever the ultimate causes of incommensurability, there is a widespread consensus among those who accept this phenomenon that its main semantic manifestation is constituted by the inability to construct a meaning-preserving translation between two languages or theories. There may, on this view, be rival theories between which the degree of conceptual disparity or meaning-variance is so radical or pervasive as to preclude any possibility of translation, whether total or partial. This property of untranslatability, and its correlate deductive disjointedness, is absolutely fundamental to the thesis of incommensurability. If meaning-variance by itself were to exhaust the entire content of the thesis, it could scarcely be said to be a critical or radical claim at all; and would hardly constitute a counter-argument to the empiricist or critical rationalist conceptions of science against which it was directed.

One need not analyse further the notoriously nebulous concept of 'meaning' in order to establish the following point. If some term R has a different meaning in T than in T', then any meaning-preserving translation from T into T' *cannot* be literal. If it were literal, it would map occurrences of R identically, and thus fail to preserve meaning. If, conversely, meanings are to be faithfully conserved, the translation will have to transform R into some expression equivalent in meaning but *ex hypothesi* syntactically different. This argument, as I remarked, is simple, indeed trivial. All the same it reveals a rather basic flaw in Balzer's explication. For it shows that, under his definition, if theories are commensurable then either (i) each of their shared terms possesses the same meaning in each theory, or else (ii) the translation that renders them commensurable fails to preserve meaning. I find both conclusions unacceptable. The first reduces incommensurability to mere meaning-variance, thereby trivialising it. And the second is grossly incompatible with how Kuhn and others have characterised (in)commensurability. Moreover nothing depends here on which components or aspects of meaning are taken to be the crucial ones.

There are various related objections that can be raised against the adequacy of (0). In the first place, the conditions imposed on Γ pertain only to the translation of *shared* terms and sentences of the two theories. But if this is the real crux of commensurability, why require that ρ be a *full* translation of *all* expressions? Why not accept partial translations, if these establish a common core of shared meanings and suffice to define logical relations between the theories? Secondly, it is worth noticing that Sneed's constraints on the ρ-relation are not logically necessary in order that a respectful translation satisfy (0). These conditions are in fact compatible with weaker kinds of semantic correspondence; for instance, where ρ is a many-valued rather than single-valued relation. Thirdly, and more damagingly, it follows from Balzer's definition that incommensurable theories could have commensurable 'parts': T and T' might be R-ρ-commensurable for some but not all $R \in \text{Symb}(\tau_0)$. This has the disastrous consequence that some incommensurable pairs of theories could become commensurable through a simple renaming of predicates: just star all symbols R in τ which are not R-ρ-commensurable, to form an alphabetical variant of T which is commensurable with T'.[18]

Things appear equally hopeless when one focuses on the characterisation of *in*commensurability. On Balzer's criteria all pairs of theories with disjoint vocabularies are incommensurable — for example, elementary geometry and real number theory, even though the standard reduction of the former to the latter is widely regarded as establishing the contrary. Furthermore, any two mutually inconsistent theories are incommensurable. For, suppose that ρ reduces T to T' and that Γ is any respectful translation. Let T be inconsistent with T' in the sense that for some τ_0-sentence φ,

$$T' \vDash \varphi, \qquad T \vDash \neg\varphi.$$

Then, since ρ and Γ satisfy (1.0) and (1.5), we must have $\Gamma(\varphi) \neq \varphi$, which violates (0)(ii). Another curious feature of the definition is that all theories having only mathematical terms in common are commensurable, providing there is some ρ and Γ that respects these terms in the appropriate way. In other words, for any two physical theories, say, sharing no descriptive terms but having a similar mathematical base, a reduction of one to the other will render them commensurable if all mathematical terms are translated literally.

These remarks indicate that, despite first appearances, Balzer has failed to assign translation any decisive role in reconstructing the phenomenon of incommensurability. Lexically disjoint theories and mutually inconsistent theories are always, according to him, incommensurable, even if there is a perfectly good translation between them. And at the other extreme, there could be physical theories that are commensurable, regardless of *how* their descriptive vocabularies are connected: *any* translation that fits together with the model-theoretic correspondence will do. This is clearly inadequate for Balzer's goal, for it means that theories enjoying a substantial conceptual overlap must, to be commensurable, fulfil what is for them a stringent demand of literal translatability, whereas theories with a meagre conceptual contact need only comply with comparatively weak criteria.

Balzer's attempts to improve on this explication and, in particular, to extend its range of application, also strike me as inconclusive. To give one example, he proposes to modify (0) so as to take account of *defined* terms. The idea is that the requirement of literal translatability should be extended to cover, besides shared primitives, also any terms introduced into either theory through an extension by definitions. However, this move seems radically to alter the spirit of the original constraints. It means, for instance, that any *interpretation* of T in T' (in any standard sense of this notion) would render them commensurable (subject to the usual condition of literality), since all terms of the interpreted theory would become defined terms in the interpreting theory. Reversing the previous verdict, geometry and real number theory would now become commensurable under any interpretation defining geometric primitives in the language of arithmetic. Once again, this shows the criteria are largely insensitive to the kind of translation available.

On the whole, Balzer's approach marks an advance in the structuralists' thinking on these matters. His inclusion of logical and linguistic elements should contribute to improving their framework and the account of science that goes with it; though it may mean abandoning some of the more extreme claims that were made on behalf of the nonstatement view. All the same, if the line I have been taking in this chapter is sound, there is little doubt that Balzer has failed to get to grips with the notion of incommensurability, and has made no concrete progress towards justifying Stegmüller's rationality thesis. The evidence against *SR* remains, in my view, incontrovertible.

It may be objected that at least some of my complaints against the structuralist criteria of commensurability could be parried by effecting suitable repairs to Balzer's explication. I accept that some of the minor details could be polished up, but it also seems to me that the basic intuition underlying (0) is faulty. Balzer and I are in complete agreement in holding that translatability is a core element of what is involved in the claim that theories are commensurable. But we diverge, seemingly irreconcilably, in our conceptions of the precise role and function of translation in this context. He insists that, among other semantic requirements for commensurability, two constraints on translation are fundamental. An intertheory translation must, for him, be a *full* mapping of all expressions, besides being homophonic or *literal*. I have argued instead that neither of these constraints on translation is necessary to ensure commensurability. I would also add that they need not be sufficient, though they may on occasion be extremely interesting and useful properties to have. As I see it, they simply fail to drive a wedge between theories that are logically and rationally comparable, and those that are not.

However, the litmus test of this view must, in the end, be subordinate to actual examples and case studies. This being so, in Chapters 6 and 7 I shall return to this problem and try to indicate in more concrete terms why translations not fulfilling Balzer's conditions may nevertheless establish logical comparability and commensurability of the kind needed for the rational appraisal of competing theories.

For the time being, we shall turn our attention to a quite different approach to the issue of commensurability.

3

RESEARCH TRADITIONS, INCOMMENSURABILITY AND SCIENTIFIC PROGRESS

1. PROBLEM-SOLVING MODELS OF SCIENCE

The view that scientific method is in large measure a problem-solving procedure has a long and impeccable pedigree. In recent times, the most notable exponent of this idea has been Karl Popper. It was he who stressed in particular the idea that theories *arise* in response to a given set of problems, a process which he described by elaborating a so-called *situational logic*. Aside from Popper, many other writers on the history and methodology of science have emphasised the importance of problems, not only for the genesis of theories, but for their development too. A prime example is of course Thomas Kuhn, who has taken problem-solving to be the principal factor guiding normal scientific research.

Problem-solving models of science have acquired even firmer footing over the last decade or so, chiefly it seems as a result of the appearance in 1977 of Larry Laudan's *Progress and its Problems*. Laudan's book had an immediate and startling impact. Deservedly so, for it is an impressive piece of writing that breaks new ground and covers old terrain in varied and novel ways. In many respects Laudan's theory of science is bold and provocative. Though assembled from a number of tried and trusted parts, its mixture is unique and its construction illuminating. It looks set to be with us for a long time to come.

Whereas Popper's treatment of problems led him to a strongly realist and anti-conventionalist view of science, Laudan's approach has much in common with instrumentalism and conventionalism. It is instrumentalist in taking problem-solving rather than truth-seeking to be the fundamental aim of scientific inquiry.[1] And like conventionalist philosophies it underlines the point that the cognitive content of a scientific theory is not exhausted by its empirical consequences but is also characterised by its conceptual structure.[2] The anti-realist element of Laudan's philosophy has to some extent emarginated him from the mainstream of current thinking about science and its methods. And, not surprisingly, critics of *Progress and its Problems* have pulled no punches on this score and prompted no less hard-hitting responses from its author.[3] Moreover, Laudan is a rebel in other respects too. He is no methodological pluralist of either weak or strong persuasion. Yet his brand of methodological monism is at odds with the orthodox view that a single notion of 'rationality' can capture the complexity of the scientific enterprise. His is a stable model of growth seen through changing patterns of rationality. He is no enemy of rationalism, but then no friend of discovery[4]; no disguised anarchist, but no dissenter from a mild and moderate relativism.

Reading Laudan can, at times, be a frustrating exercise. He rarely offers his opponents a fair bite at the cherry, often simplifying the subtleties of their views and leaving stones unturned and dark corners unswept. And, as many were quick to observe, his own theory at times emerges as little more than a skeleton, waiting to be fleshed out and brought to life. Here I am alluding especially to his central notion of *problem-solving effectiveness*, or PSE for short. How in practice should weights be assigned to the anomalies and empirical problems that challenge a theory or research tradition? How, in a nutshell, is an honest index of scientific progress to be achieved?

If there are obvious shortcomings in Laudan's theory, there is

also much to commend it. The historian and the sociologist of
science should find the research tradition a more malleable and
manageable creation than the Kuhnian paradigm. The method-
ologist should welcome Laudan's generous dosage of historical
source material, even while possibly lamenting the absence of any
logical and conceptual hardware. And I venture that many will
applaud the attempt at a thorough-going nonrealist account of
scientific development, even when quarrelling with Laudan's own
'confutation' of convergent realism.

I shall not embark here on a minute exposition of Laudan's
theory. The details of it have been must discussed and well
digested by now, and I shall assume that the reader is broadly
familiar with its main lines.[5] I shall therefore recall only some of
the salient points of his account, as and when they appear
relevant to the topic at hand. What I do want to consider at some
length is the novel and radical solution that Laudan proposes to
the problem of incommensurability. It represents a quite different
method of attacking this issue than those I have so far dealt with;
and it is, I believe, equally flawed. To my knowledge a coherent
criticism of this aspect of Laudan's theory has not yet surfaced
from the abundant discussion that can be found in the literature.
Aside from the fact that Laudan's treatment of commensurability
questions is captivating in its own right, there is a wider motive
for attending to it. The inadequacies of his account threaten to
endanger the success of the problem-solving model in providing a
rational picture of scientific progress. Bringing these problems
into the open might, I hope, help to clear the air, forestall this
threat and, ultimately, remove the danger.

2. LAUDAN ON INCOMMENSURABILITY

Laudan advances two general theses about incommensurability.
the first states quite simply that rival theories, or research

traditions, are not as a matter of fact incommensurable since they always share a common store of empirical problems. The second claims that even if one were to concede the incommensurability of a pair of research traditions, cognitive appraisal of them and rational choice between them would still be feasible. So even in this extreme case the question of scientific progress could be raised and in principle settled.

Though neither of these claims is precisely formulated, it is at once clear that they are quite different in character. The former relates to the way scientific theories are or have been in the past; and is designed to be a straightforward rebuttal of the doctrine of incommensurability. In contrast, Laudan's second thesis really amounts to a logical or conceptual claim concerning the nature of his own model of scientific growth. It functions as a kind of fallback or failsafe argument intended to dilute the problem of incommensurability rather than dissolve it.

In due course I shall scrutinise both claims in some detail. But before examining the merits of each of them independently, it is important to notice that something of a general dilemma seems to face any attempt to force through both theses simultaneously. The problem, briefly, is this. Once one agrees to take the doctrine of incommensurability, in some form or other, at all seriously, it is likely one will acknowledge that there are indeed pairs of scientific theories (or even whole research traditions) that are at least plausible *candidates* for incommensurability. Whether they are ultimately regarded as such, depends in the last resort on what sorts of criteria of incommensurability are adopted, what interpretation is put on the theories, and what standards of theory comparison are operative.

Now, if one's goal is, like Laudan, rationally to reconstruct scientific development within a single type of explanatory model (say the problem-solving one), one is liable to notice a tension that arises from two opposing directions. On the one hand, one could succumb to 'weak' criteria of incommensurability, choosing a construal that risks admitting many positive instances of incommensurability among scientific theories. But then, by the same token, one might reasonably expect to retreat to a position like that expressed by Laudan's second claim: one's model ought then to be rich enough to cope with the rational appraisal of theories notwithstanding their incommensurability. For example, the discussion in earlier chapters made quite clear that if incommensurability were defined as *mere* conceptual disparity or meaning-variance, one would expect to have little trouble in handling cognitive theory assessment by means of comprehensive techniques of translation.

On the other hand, one could plump for severe standards of incommensurability, hoping thereby to avert the likelihood that any genuine case of this phenomenon actually emerges. But this option greatly reduces the possibility of falling back to the second type of claim: rational evaluation in spite of incommensurability. For, if one's criteria are such as to render virtually all rival pairs of theories commensurable, the cognitive comparison of any that escape the net is bound to be problematic.

The dilemma just described confronts Laudan's theory in the following way. He develops a model of rational scientific progress based on a complex and highly *relativised* notion of problem-solving effectiveness, PSE. It thrives on the element of *competition* between rival research traditions; and it relies thus and heavily so on the idea of *shared* empirical problems. Now, providing one can legitimately appeal to the presence of shared problems, one has good *prima facie* grounds to argue for commensurability. But as soon as one ceases to rely on problem sharing, not only incommensurability results, the whole process of comparing the PSE's of different theories is undermined and the model of rational progress breaks down. Part of the difficulty plainly arises from trying to defend both types of thesis in the same breath. But what is even more worrying in Laudan's case is that neither of his arguments looks in any way convincing in its own right. I shall now turn to the reasons.

Laudan's assertion that there are no pairs of incommensurable theories in science is not backed up by any solid argument at all. He examines no case studies, nor does he offer up any grounds *a priori* on the basis, say, of a suitable theory of meaning. One is even at a loss to find evidence for the plausibility of his claim, because he actually settles for severe standards of commensurability which will, in practice, be hard to match.

One aspect of his argument consists of a challenge to the chief proponents of the doctrine of incommensurability: Kuhn and Feyerabend. But no advantage is gained from this, because the challenge itself is inaccurate and confused. Kuhn and Feyerabend are moulded by Laudan into a single figure which represents neither of them individually, nor what is common to each. The two authors are alleged to subscribe jointly to the view that incommensurability results from a conceptual disparity that is so deep as to forbid any kind of logical contact between the theories concerned, including translation and deduction. And they are held to infer from this that no rational comparison of such theories is therefore possible. But the first of these views does not reflect the truth about Kuhn,[6] and the second has been explicitly denied by both Feyerabend and Kuhn.[7] Even if one reads Laudan as interpreting rather than reporting the content of their writings, it is incorrect to assimilate their accounts in the casual manner which Laudan adopts.

Even apart from this, there is no attempt from Laudan's side to relate his own concept of commensurability with those of Kuhn and Feyerabend. And, consequently, it is unclear, without further evidence, why his criticism of them should count in favour of *his* thesis of commensurability, or, conversely, why his own arguments for commensurability should be taken to contradict the claims of Kuhn and Feyerabend.

For Laudan, incommensurability can only arise through the

absence of shared problems. The burden of justifying his claim for commensurability rests therefore on the notion of 'empirical problem' and its status in different theories. But 'problem' is, in Laudan's account, a primitive and unanalysed term. It is sometimes understood generically, as in the phrase 'the problem of free fall', and it sometimes gets a more specific interpretation, as in the case of 'the law of light refraction in a given medium'. Conceptual problems are always, and empirical problems almost always, regarded as theory dependent.

Laudan does not, as a rule, make any special distinction between problems and their formulations. In principle, therefore, one could construe problems propositionally, or take them in some other, nonpropositional light. The second option might have some appeal for the defender of commensurability who wished to argue that two theories unrelatable by any logical or conceptual link might still address a common problem. But this is not Laudan's strategy. For one thing, he offers no clue for how to re-identify a problem across 'incommensurable' formulations of it. For another, it is the manner in which problems are expressed and solved that affects the progressiveness of a theory (or research tradition) and determines its cognitive relation to a rival. And on the matter of problem *solving* Laudan is quite clear: a theory *T* solves a given problem if "*T* functions (significantly) in any schema of inference whose conclusion is a statement of the problem" (Laudan, 1977, p. 25).

If one sticks to the linguistic reading of problems as *statements* or *propositions*, then Laudan's view amounts to this: Rival theories share at least some descriptive terms or expressions (potential problem solutions) which, though not purely observational in the sense of belonging to some theory-neutral language, nevertheless form part of a language which can be fully understood independently of the specific postulates of the theories in question. As he puts it,

> *. . . with respect to any two research traditions (or theories) in any field of science, the[r]e are some joint problems which can be formulated so as to presuppose nothing which is syntactically dependent upon the specific research traditions being compared.*   (Laudan, 1977, p. 144)

Plainly, Laudan is admitting here both the general context dependence of problems and problem-solutions and yet claiming the *relative* independence of *some* shared problems from the specific theoretical frameworks in which they are raised and solved. This is the crux of his case for commensurability.

One of the difficulties with Laudan's account is that he does not distinguish and classify different problem types according to their grades of 'theoreticity'. Among empirical problems, therefore, one should presumably include: (i) very general and low-level problems which can be *prima facie* formulated in 'ordinary language' independently of any theories whatsoever; (ii) problems concerning observable entities and events which require theoretical presuppositions for their formulation, but which are 'non-theoretical' in the context of the theory which solves them; (iii) fully 'theoretical' problems which arise solely in the context of a particular theory.

It seems fair, then, to suppose that Laudan's thesis pertains to problems of the second kind. It would be trivially false about type (iii) problems. And, if asserted about type (i) problems the thesis would seem to be devoid of any substantial content. Moreover, in this case it would scarcely be an effective rejoinder to the problem of incommensurability. For, even if one could show by examples that research traditions always share common problems of the first type, this would serve only to identify them in some intuitive and pragmatic sense as rivals. Since, for Laudan, empirical problems are generally context dependent, such a claim would not rebut the semantical thesis of incommensurability, since it would not follow without further argument that the respective solutions offered by different research traditions to such low-level problems could be brought into logical relations of mutual entailment, inconsistency and the like, and thus properly recognised as compatible or conflicting solutions to the same problems.[8]

If this is a correct diagnosis of Laudan's position, he seems then to be asserting the existence of what Hempel has called an *antecedently understood* vocabulary and what Sneed has designated a set of *nontheoretical* terms. As we saw in Chapter 1, Sneed's view is that for any scientific theory T, some of its terms (T-theoretical concepts) can be held to be interpreted in a T-dependent way, whilst others can be counted as nontheoretical *with respect to T*, though they may of course be theoretical in the context of another theory T'.

This part of Laudan's claim, is relatively uncontroversial. But if one now spells out the full extent of his argument, one sees that in effect much more is involved. For, his underlying assumptions proceed as follows. Let T_1 and T_2 be rival theories within distinct research traditions, R_1 and R_2 respectively; and suppose that S_1 and S_2 are the sets of nonlogical concepts which are T_1-non-theoretical and T_2-nontheoretical in that order. Now, Laudan's first requirement is that $S = S_1 \cap S_2$ is a nonempty set of terms in the language of which empirical problems common to T_1 and T_2 can be expressed. And his second claim is that all members of S are T-nontheoretical *for any theory T in either R_1 or R_2*.[9]

This is a bold hypothesis which not only exceeds by an ample margin what is usually taken to be necessary for commensurability but is just the kind of assertion that most rationalist critics of the incommensurability thesis have long ago abandoned. Consider the second part first. In order to establish that T_1 and T_2 deal with a common problem, one need only show that terms in S have the same meaning in T_1 as they have in T_2, or, more generally, that any differences in their meanings can be suitably corrected for. It may sometimes be sufficient, but it is certainly never necessary, to know that they are interpretable independently of *any other* theories in R_1 and R_2. This observation applies irrespective of any particular theory of meaning, and it holds whether extensional, intensional, or any other component of meaning is specified. Moreover, contrary to Laudan, I would suggest that it is the rule rather than the exception that research traditions contain the theories which characterise and interpret the facts or problems besides the theories which explain or solve them. Take the Newtonian research tradition as an example; it clearly embodies both basic theories about the geometry of space and time, as well as the kinematical, mechanical and cosmological theories which explain properties of, and physical phenomena in, space and time.

The first part of Laudan's claim is equally redundant for the purpose of saving commensurability. Two theories might formulate a problem in quite distinct ways and even use different vocabularies in the process. It follows that if commensurability is to be a matter of problem-sharing, it need not, and indeed should not, be dependent on concept-sharing. And, conversely, the presence of shared terms does not guarantee the existence of joint problems. Translation of some sort will generally be required if commensurability is to be argued, rather than taken for granted. It is no virtue of Laudan's model that it avoids questions of intertheoretic translation by ignoring them.

What I am suggesting here is not that the shared problem cannot function as an arbiter of commensurability, nor even that Laudan's commensurability thesis must be false. The problem is rather that his argument, such as it is, seems to embody claims which go way beyond what is really necessary to defend commensurability in his own terms, and to rest on assumptions that are now largely discredited. One might be prepared to waive this objection if Laudan's second thesis could be sustained; for there would then be less reason to trouble over incommensurable theories anyway. But the second thesis is less cogent than its author pretends, as we shall soon see.

3. LAUDAN'S SECOND THESIS

Laudan's idea that even incommensurable research traditions can be objectively evaluated and compared is articulated in the following passage.

> Now, an approximate determination of the effectiveness of a research tradition can be made *within* the research tradition itself, without reference to any other research tradition. We simply ask whether a research tradition has solved the problems which it set for itself; we ask whether, in the process, it generated any empirical anomalies or conceptual problems. We ask whether, in the course of time, it has managed to expand its domain of explained problems and to minimize the number and importance of its remaining conceptual problems and anomalies. In this way, we can come up with a characterization of the progressiveness (or regressiveness) of the research tradition. . . . It is thus possible, at least in principle and perhaps eventually in practice, to be able to compare the progressiveness of different research traditions, *even if those research traditions are utterly incommensurable in terms of the substantive claims they make about the world*!
> (Laudan, 1977, pp. 145—146)

One point here that needs special emphasis is acknowledged by Laudan in a footnote. In the circumstances described above, when one assesses the progressiveness of a research tradition, one's attention can only be confined to the problems and anomalies that are expressed *in its own terms*. Problems pertaining to a rival, and by hypothesis incommensurable, research tradition play no part at all. Each tradition is thus evaluated entirely on its own merits, without reference to the problems, anomalies, concepts, methods and standards of appraisal present in other research traditions.

In some cases it may well be feasible to undertake the task

Laudan envisages. But for several reasons it falls short of achieving the desired end. One difficulty that arises concerns the problem of choosing between incommensurable research traditions. But by far the worst drawback is that the heuristic force of the problem-solving model is severely curtailed. In fact, in a certain sense, the model breaks down completely.

Laudan distinguishes in general the two pragmatic contexts of *choice* and *pursuit*. For him, one may find rational grounds to pursue a research tradition even if one refrains from taking the stronger stance of *accepting* it. Though each of these epistemic attitudes is possible, it is clear that, for Laudan, rationality is most intimately linked to the choice context: it is always rational to choose the most progressive research tradition available. But in the case of incommensurability, the whole issue of choice is a murky one. If two research traditions have no problems in common, it is, first of all, unclear in what sense they may be considered rivals, and secondly, it is doubtful whether there could be any premium set on choosing between them. At least the motivation for having to make a rational preference here would have to be more fully explained.

The situation characterised here also robs Laudan's theory of much if not all of its heuristic power. In general, to assess the progressiveness of any research tradition one must calculate the PSE of its constituent theories at a given time. These measures, in turn, depend on the weights assigned to the various problems and anomalies which the theories confront. But weights are not independent of the parent research tradition; they may differ between one tradition and a rival. So, it could be argued that the PSE of these theories — hence ultimately an evaluation of the progressiveness of the tradition itself — can always be ascertained from *within that same tradition,* and conclude that this results in a biased and even circular index of rationality.[10]

To prevent this objection from undercutting Laudan's theory, one must assume that rival research traditions display a large number of shared problems. Scientific progress can then be seen to be achieved through the interaction between different theories from rival traditions in their efforts to solve those problems. Laudan agrees:

> ... it is basically the shared empirical problems which establish the important connections between successive research traditions; it is these, and these alone, which must be preserved if science is to exhibit that (partially) cumulative character which is so striking about much of its history.   (Laudan, 1977, p. 140)

Not only cumulativity is at stake, but the very possibility of rational assessment. Presumably it is because at least some problems are equally important and some anomalies equally irksome to rival research traditions that a fair measure of their relative progressiveness can be determined. A rational comparison can be expected whenever a research tradition pays attention to the failures and gives credit to the accomplishments of its competitors. But this situation clearly prevails only as long as rival traditions survive on a healthy diet of shared problems and common methods of evaluation.[11] When incommensurability arises, this kind of common ground is lacking, and the basis for rational choice under the terms of Laudan's model is lost.

Laudan himself stresses the highly relativised character of progressiveness in general:

> All evaluations of research traditions and theories must be made *within a comparative context*. What matters is not, in some absolute sense, how effective or progressive a tradition or theory is, but, rather, how its effectiveness or progressiveness compares with its competitors.   (Laudan, 1977, p. 120)

It should be obvious by now, however, that the assumption of incommensurability destroys the capacity of Laudan's theory to serve up impartial judgements of relative progressiveness, and hence to provide a cognitive index of scientific growth.

My objection can also be strengthened somewhat. Imagine a situation in which some established research tradition is challenged by the arrival of a new theory *T*. Suppose also that *T* is incommensurable with all the latest theories in the existing tradition, and is actually accompanied by an emergent, rival, incommensurable and potentially revolutionary research tradition. For Laudan, it may be rational to pursue the new tradition if it shows a higher (momentary) rate of progress than its rivals. But by what criteria? Generally, says Laudan, there are three: the contemporary competitors of *T*, the "prevailing doctrines of theory assessment", and the previous theories within the research tradition, must be taken into account.[12] Now, under the present assumptions, the last of these categories is empty, and the first becomes, as I argued, irrelevant under the hypthesis of incommensurability and the absence of shared problems. Moreover, even the second category may be inapplicable, if, as is likely, *T* is associated with new methods and standards of appraisal. It is these and these alone by whose lights *T* must be judged.

It is hard to dodge the conclusion, then, that the initial rate of progress of a theory like *T* cannot be ascertained in an objective manner that makes good sense as a comparative yardstick. The problems *T* solves are of its own making, the adequacy of its solutions is to be judged by its own standards, and, more pertinently, *it cannot generate any anomalies* as Laudan seems to think, because for this one would require the existence of another theory to have successfully solved a problem in *T*'s domain; a state of affairs forbidden by the assumption of incommensurability.

Suppose, nevertheless, that some grounds are found for pursuing *T*, and that sooner or later its proponents detect inadequacies in some of the theory's explanations or problem solutions.

Let us even imagine that in its own terms T is 'refuted'. According to a widespread consensus to which Laudan also subscribes, it is always possible in this case to modify T so as to "save the phenomena"; for instance, to change the interpretative rules of T so that the refuting instance is removed from its domain. Under most methodological theories such a strategy would normally count as *ad hoc*, and the revised theory, T' say, would be regarded as less progressive. But Laudan has a novel response to the problem of *ad hoc*ness. He rejects all syntactic or formal criteria of *ad hoc*ness and admits only the device of PSE. For him, genuine *ad hoc*ness arises only if the new theory exhibits a *decrease* in PSE. Other things being equal, if T' generates no new conceptual problems and loses no problem-solving ability elsewhere, then T' at once shows a greater PSE than T; its adoption or pursuit amounts to a progressive step.

This counter-intuitive situation cannot even be corrected by arguing that probably a third theory will emerge and render T' regressive by resolving the 'refutation' without a corresponding shrinkage of domain. At least, this possibility cannot arise if one assumes that T and T' are the only theories in the research tradition and are incommensurable with their contemporaries. It is plain, therefore, that Laudan's model is unable to legislate against a new research tradition that makes itself impregnable by adopting sloppy standards and accommodating all 'falsifications' by means of *ad hoc* adjustments and 'exception-barring' methods.

4. PROGRESS AND THE PROBLEM-SOLVING MODEL

Is there any way to escape the conclusion that the problem-solving model falters and founders under the pressure of the doctrine of incommensurability? I can see no compelling grounds for despair, but it is evident that Laudan's own account will have to be corrected and supplemented at the appropriate places. With the example just described in mind, one might try to argue that research traditions always embody at least some explicit criteria of 'refutability' (and methodological principles in general) that are relatively stable or even time-invariant, and that some such norms of appraisal are shared across all the research traditions in a given field of science. It is not clear that Laudan himself would approve such a move, since, in more recent writings (e.g. Laudan, 1984), he underscores the need in general to recognise methodological disputes in science, and to analyse the means of resolving them at a higher, 'axiological' level.

Another course of action would be to re-introduce formal standards of *ad hoc*ness alongside Laudan's epistemic ones, with the aim of showing that formally *ad hoc* theories are always regressive in that they generate conceptual problems and anomalies. But the most promising strategy, I believe, is to return to the fundamental issues of meaning change, translatability, problem-sharing, and so forth, that are central to the commensurability and rationality debate. By equipping the problem-solving model with the right kind of analytical tools and conceptual hardware, one should be able to build a more plausible defence of Laudan's first thesis. And, in later chapters, I shall be looking at some of the ways of achieving this goal.

But, before moving on from Laudan's model of science, I want to make a last attempt to see, in a more concrete case, whether anything of his second thesis can be salvaged. Could it not happen that, in practice at least, the 'internal' PSE-count of a research tradition suffices, for most purposes, as a cognitive index for rational choice? The short answer is: *No*. And, in fact, one can argue that not only are numerical measures of PSE indecisive in this respect, even the qualitative *direction* of progress may exert less influence than one usually imagines.

Suppose, then, that one wishes to apply the criterion of problem-solving effectiveness to a pair of established, and possibly incommensurable, research traditions. To fix ideas a little more precisely, let us take as examples the classical or Platonist, versus the intuitionist, research tradition in the foundations of logic and mathematics. For present purposes, one need not take a definite stand on the question whether these two traditions are commensurable or not. What is important is, first, that there is a clear, pragmatic sense in which they are rivals; secondly, that some sort of case — not necessarily a watertight one — can be made for their incommensurability.

As far as the second point is concerned, for example, the following factors seem especially relevant. As compared with Platonists, intuitionists adopt a different construal of mathematical objects (conceptualism). They employ alternative concepts and methods for theorem-proving (choice sequences, spreads, Kripke models, etc.) and for evaluating the cogency of proofs (constructivism). And, besides this, intuitionists employ nonclassical logics and reject many problems of classical mathematics as meaningless. Individual theories belonging to the two research traditions can, of course, be logically compared, for instance by the method of translation and interpretation. But, though this comparison proceeds relatively smoothly from the *classical* standpoint, from the other side the matter is less straightforward, due to differences in metatheoretical assumptions and to the special role of the creative subject in intuitionism. In short, whether one views these traditions as truth-seeking, or, alternatively, as problem-oriented, significant conceptual, ontological and methodological differences between them can be detected.[13]

For each of these research traditions, one can try to distinguish a range of empirical and conceptual problems, and perhaps even anomalies.[14] Consequently, at least in principle, one could carry out an internal assessment of their progressiveness at any given time. On the assumption of incommensurability, a straight numerical comparison of their PSE, would, as I already argued, be of little effect. But, suppose that over a fair period of time some marked qualitative differences could be observed. Suppose that one of these traditions happened to be, in its own terms, highly progressive during a period in which the other was, in *its* terms, stagnant or even regressive. Would one expect to see a rush of Platonists turning constructivist, or *vice versa*? Would it be rational for mathematicians to switch allegiance in these circumstances?

It is reasonable, I think, to expect a negative answer to both these questions. Naturally, to come to an informed judgement one would need to have all the relevant facts at hand. What alternative theories or research traditions are available, for instance? What factors are affecting in each case the calculation of PSE? There may indeed exist real or hypothetical situations in which the pursuit of a mathematical research tradition at a given time would be irrational. Examples would be: if some accepted metamathematical results showed certain background assumptions of the tradition to be false, or certain proclaimed goals to be unattainable. One thinks here of the situation faced by the Hilbert Programme around the time of Gödel's incompleteness results.

On the other hand, it is apparent that research traditions in mathematics, as elsewhere in science, do evolve and adapt to changing circumstances, not only by developing new theories but also by modifying or even abandoning some of their background presuppositions. Formalism, as a foundational doctrine, did not expire as a consequence of the incompleteness theorems. And even substantial parts of Hilbert's Programme survived through the work of Gentzen and others. Thus, whilst one might well expect to see research traditions of this kind move in new directions and sometimes relinquish certain problem-areas previously thought to be fruitful, it is hard to imagine a situation in which the proponents of one tradition would unanimously abandon it in favour of a more progressive rival or rivals. If that is the case, there is more to allegiance than a commitment to problem-solving effectiveness.

If correct, this last conclusion might be damaging, not only to Laudan's analysis of commensurability questions, but to his theory of progress in general. If loyalty to a research tradition is taken to be a not-entirely-rational affair, one ought to infer that scientists sometimes — and perhaps even often — fail to act rationally, and that Laudan's theory explains only a portion of the cognitive development of science, namely that part clearly portrayable as the pursuit of progress in the problem-solving sense. It would seem that to preserve the idea that science is basically a rational enterprise, whilst maintaining one's trust in PSE as an adequate index of progress, one must acknowledge that progressiveness cannot be the sole criterion of rationality.

It is important to remember here that, for Laudan, progress is evaluated for a research tradition on the basis of the PSE of its component theories *and on nothing else*. For example, the background assumptions of a tradition play an indirect part in the determination of PSE, since they may influence the weights assigned to problems and anomalies. But they have no *direct* bearing on the question of rationality, because they cannot count as separate and independent grounds for pursuing, altering or abandoning a research tradition. Now, to give up the identification of rationality with progressiveness in the problem-solving sense would be to give up one of the fundamental tenets of Laudan's theory. Perhaps a viable problem-solving model of science could be developed without this central assumption, but it would not be Laudan's model. However, the assumption can, I believe, be retained, and an account of rational scientific progress still be delivered, providing that two further conditions are met. One is that we obtain a more precise grip than Laudan achieves on the internal structure of the research tradition. The second condition, which I hinted at a little earlier, is that we develop a systematic analysis of intertheoretic relations that can be used to study the cognitive connections between rival theories and to establish, where possible, their commensurability. In subsequent chapters I shall be exploring both these ideas in more detail.

4

THE LOGIC OF REDUCIBILITY

In science, as well as in philosophy, the method of reduction is a fundamental tool for reconciling *prima facie* conflicting theories and for rendering apparently disparate conceptual frameworks rationally comparable. As a field of systematic investigation by philosophers of science, reduction has flourished at least since the classical analysis provided by Ernest Nagel (1949).[1] From that time on, the subject has been intensively studied, from a variety of philosophical perspectives, using a variety of logical and mathematical methods, and taking a variety of standard examples as paradigmatic.

In the philosophical literature on reduction, emphasis has largely, if not exclusively, been centred on syntactic, derivational features of this relation; the influence of Nagel's deductive-explanatory model of reduction being decisive. This way of approaching the subject has tended to focus attention on certain epistemological *types* of reduction, and on particular philosophical issues they foster, to the exclusion of others. As examples one might mention the attention paid to the distinction between homogeneous and inhomogeneous reduction, to the status of bridge laws, to the problem of supervenience, and so on.

So far in this book I have been discussing various aspects of reduction that arise naturally in connection with problems of intertheory translation and commensurability. It is now time to attack this issue more directly. I shall propose a uniform treatment of reduction, or more accurately a system of reductive types of intertheory relations, within a general model-theoretic framework. The approach I adopt has been developed in collaboration

90

with Veikko Rantala, and various aspects and applications of it have already been sketched in our joint publications.[2] Though in several respects our model of reduction in science is a straightforward generalisation of some of the traditional accounts, it leads nonetheless to a somewhat heterodox classification of reductive relations that serves to highlight certain features of reduction which tend to be obscured by the standard treatments.

Before getting down to the fine details of the formal model, one should first consider how different kinds of reducibility may occur in science.

1. TYPES OF REDUCTION

In the context of scientific languages and theories, the term 'reduction' occurs in a plurality of senses. A reduction may, for instance, be said to be *conceptual* when it provides for the analysis of some scientific concept on the basis of other concepts that are in some way more basic or fundamental. A standard example is the reduction of temperature (of a gas) to mean kinetic energy (of the gas molecules). Analogously, one can say that an *ontological* reduction has been accomplished when certain entities or systems of entities described within a theory can be shown to be composed of, defined from, or constructed out of, more basic entities or collections, possibly described within some other language or theory. The reduction of geometric notions like points, lines and planes, to numbers, and the further reduction of numbers to sets, count as typical examples.

In either case, reduction is not merely a device for ordering and systematising one's conceptual scheme, or a means to satisfy a philosophical taste for ontological economy; though it may achieve both these ends. First and foremost, it is a method for enhancing scientific understanding and increasing our ability to explain, predict and control. One aspect of this process might be termed *explanatory* reduction. It can be said to occur when the problems solved and explanations proffered within one framework or theory can be reduced to problems and explanations handled by another framework or theory. One might, in this way, solve a thermodynamical problem by re-expressing it in the language of statistical mechanics, or answer a geometrical problem using analytical methods.

A further type of reduction is also explanatory, but to distinguish it from the former I shall call it *theoretical* reduction. This name can be used to cover those cases where a scientific principle, a theory, or even an entire branch of science, is reduced to a more fundamental principle, theory or science. What are explained here are therefore not (sets of) problems, but rather (sets of) laws.

Plainly, there is an intimate connection between these two types of reduction; and indeed most philosophers have not troubled to distinguish them at all.[3] But a closer inspection reveals that, from a logical point of view, a clear division can be carved out. Suppose that one has a purported reduction of a theory T to a theory T'. In the simplest possible case, one can interpret this relation to be of the 'theoretical' kind when all laws of T can be derived from the (conjunction of the) laws of T'. Hence, if the theories are as usual deductively closed, one knows that any logical consequence of T is likewise a logical consequence of T'. On the other hand, the simplest kind of 'explanatory' reduction relation would tell one that the truth in T' of any given statement in the language of T is sufficient to ensure its truth in T. Thus, on this reading, an 'explanatory' reduction might not be 'theoretical', if some valid statement in T happened to be undecidable in T'; and a 'theoretical' reduction need not be 'explanatory', if some undecidable statement of T turned out to be derivable from T'.

There is another way to grasp the difference which is somewhat closer to actual practice. Consider the manner in which explanations and problem solutions are effected within a certain scientific theory T. One usually requires, besides the laws of T, also extra premises: initial and boundary conditions, and so forth. For the problem solution to be adequate these additional assumptions must at least be compatible with the original theory T. But from this it does not follow automatically that the same assumptions will be compatible with a theory T' from which the laws of T are derivable. Consequently, if T' is strictly stronger than T, there may be potentially admissible explanations by T that are excluded by T'. For this reason, then, one has to regard the transfer of explanations, problem-solutions and the like, from one theory to another, as a feature different from, and perhaps even independent of, the explanation of a law or theory.

Traditionally, the derivability of laws has been viewed as the touchstone of scientific reduction. Coupled with a marked tendency to construe laws as statements, or propositions in statement form, this has produced predominantly syntactical models emphasising what I have here termed theoretical reduction. Taking their cue from Nagel's classic analysis, most writers have thus endorsed some version of the idea that T is reduced to T' when the laws of T', possible conjoined with special assumptions and connecting hypotheses (bridge laws), logically entail the laws of T. In this way, aspects of *conceptual* reduction have also been underscored; the bridge laws — especially when in the form of biconditionals — serving up the relevant conceptual links.

As we saw in Chapter 1, however, structural, or broadly speaking model-theoretic, alternatives to the syntactical model of reduction surfaced early on in the work of Suppes and his followers. And in contemporary reduction studies several different structurally-oriented accounts are available.[4] These are, for the most part, more sophisticated and mature than the naive but better-known derivational models of reduction that still dominate the philosophical literature. As regards basic intuitions, however, the two approaches have a good deal in common. The chief difference seems to be that most writers in the structural tradition have taken explanatory rather than theoretical reduction as their explicandum. Sneed, for instance, includes among his informal constraints the condition

> If S' is a statement of T' about a certain physical system and S is a statement of T about a *corresponding* physical system, then S' is true only if S is true.   (Sneed, 1971, p. 220)[5]

And Stegmüller (1973/76) echoes a similar sentiment.

In terms of structures, recall from Chapter 1 Suppes' requirement that for *every* model of the reduced theory T, a model isomorphic to it can be recovered (constructed from) some model of the reducing theory T'. Suppose that ρ is the mapping that associates in this way T-models with T'-models, and that correspondence between sentences is determined by a translation map Γ which respects ρ in the sense of (1.5); in other words, for all models $\mathfrak{M}$ and sentences φ,

$$\rho(\mathfrak{M}) \vDash \varphi \Leftrightarrow \mathfrak{M} \vDash \Gamma(\varphi).$$

Then, under suitable assumptions, Suppes' condition entails that, for any sentence φ in the language of T, if the translation of φ is a consequence of T' then φ itself is a consequence of T. In short, at least on one plausible reading it complies with Sneed's desideratum quoted above. And the same conclusion can be maintained for the revised explication of reduction introduced by Mayr (1976).[6] The Suppes' type of analysis thus seems to favour explanatory over theoretical reduction, and the model-theoretic approach in general lends itself to a simple characterisation of the distinction between these two varieties. Plainly, the former demands that (up to isomorphism) *all* models of the reduced theory possess counterparts among models of the reducing theory, whereas the latter seems to require that *every* model of the reducing theory determine some model of the reduced theory.

Some philosophers have also made use of a model-theoretic framework to define a weaker conception of reduction, applying to those cases where there is merely a uniform determination of T-models from T'-models. This type of relation has been studied in the philosophy of mind, where it is associated with the thesis of *physicalist materialism* or *determinationism*: nonphysical facts and phenomena (described by T) are in this case said to be determined by physical facts and phenomena (described by T').[7]

Whilst conceptual reduction can be neatly represented in the familiar syntactic approach, the structural perspective takes onto-logical reduction in its stride. It does so because the relevant entities and processes (or suitable surrogates for them) are present in the correlated models. By employing generalised kinds of definability (e.g. of the Gaifman or Hodges type discussed in Chapter 2), one can therefore characterise the way in which the ontology of one theory is reduced to the ontology of another. This division of labour, though suggestive, does not however mark out sovereign territories: in the last resort the model-theoretic account has no more a monopoly on ontological reduction than does the syntactic view on conceptual reduction.

2. GENERALISATIONS

Reduction, then, comes in various guises and is susceptible to both syntactic and semantic description. To what extent could one get by with only one of these descriptions and forgo the other? The answer, I believe, is that both are essential for an adequate theory of reduction. Suppose, as structuralists do, that one takes the models as basic. We saw in Chapter 1 that under certain plausible assumptions one can then infer from the model-theo-retic correlation an appropriate linguistic connection, too. But this fact does not license one to omit the syntactic side of the relation. For one thing, those ancillary assumptions would have to be explicitly set forth in the reconstruction. But, more significantly, there would even then remain two degrees of freedom to be tightened up. One would have to fix an appropriate *language* for each of the theories concerned, and, secondly, settle on one of the possibly many different translations which respect the chosen model-theoretic correspondence.

Many philosophers, on the other hand, have been content to rest with the linguistic perspective. For them, presumably, semantic structures would appear, if at all, at a secondary stage and as 'derived' entities. But this strategy is equally myopic, for there is no unique passage from syntax to semantics. First it has to be decided what sort of semantic interpretation is to be assigned to the theories concerned, in short: what *kinds* of models are to be considered.[8] And, furthermore, a functional type of semantic correspondence will be assured only given certain additional facts about the syntactic connection between the languages in question. Lastly, in many of the more advanced mathematical sciences, especially theoretical physics, scientists frequently formulate their theories *around* a certain abstract structure, like a Hilbert space or a particular group. In such cases it is not only more natural, it may even be indispensible, to introduce into the semantic interpretation the appropriate mathematical structures from the outset.[9]

It is well known that the standard deductive-explanatory model of reduction has not gone unchallenged or survived intact. Not surprisingly, it was one of the prime targets of the mounting antipositivist attacks of the early 1960s, and in Chapter 6 below I shall consider Feyerabend's critique of it in a little more detail. For present purposes, it suffices to remark that the most important direction of change undergone by the standard model of reduction was occasioned by what one may term the problems of incompatibility and approximations. These problems arise most prominently in cases where a supplanted theory is held to be reducible to the theory that supersedes it. In relating respectively

Kepler's laws of planetary motion and Galileo's law of free fall to Newtonian mechanics, for example, there is an obvious element of incompatibility between the earlier theories and the later one, and a need to introduce idealising and approximating assumptions in order to obtain something resembling a deductive relation between them. As a consequence, a number of philosophers was persuaded simply to exclude such cases from the proper domain of intertheory reduction construed as a deductive and explanatory relation. Others have clung to Nagel's original conception with modifications to account for explanations of an *approximative* sort. These usually involve introducing not only a new logical schema of reducibility, but a new epistemological category of explanations as well.[10]

It seems plain enough that one needs an enlarged framework of intertheory relations in order to handle limiting and approximative types of reduction. From this it does not follow, however, that the different kinds of reducibility just surveyed cannot be represented in a uniform manner by means of a combination of syntactic and semantic conditions. Viewed in terms of structures, approximative reductions will have to embody some idea of 'closeness' between models of the theories concerned. And, in typical cases, one can expect such proximity to be a feature of only *some* of the models in question; for instance, those in which a given parameter approaches some limiting value (there is usually some aspect of *convergence* involved here). Other things being equal, one is still dealing basically with a correlation between models, and it is still to be hoped that a suitable kind of logical, linguistic connection will result from this.

Let us agree to say that there is a *correspondence* of a theory T to a theory T', when there is a fixed mapping F which associates to each member $\mathfrak{M}$ of a given class of T'-models a model $F(\mathfrak{M})$ of T, together with a mapping I, defined on sentences in the language of T, which takes each such sentence φ to some sentence $I(\varphi)$ in the language of T', such that I respects F. ($I(\varphi)$ is to be called the *translation* of φ).[11] In the light of the foregoing discussion, we can therefore suppose thay any reduction of T to T' can be represented by a correspondence of some kind. In other words, we can make use of correspondence as a core intertheoretic relation on the basis of which different types of reduction are subsequently analysable.

To sum up, correspondence between theories incorporates in the most general form the two syntactic and semantic constraints that are essential to reducibility: translation, at the level of language, and structural correlation, at the level of models.[12] Moreover, these two constraints interlock in the manner which has by now become familiar, namely, the translation must respect the semantic mapping.

3. RECONSTRUCTIONS

We have arrived, then, at some of the core features of reduction, together with a rough classification of reduction types. Let us now try to convert these into a more exact schema of intertheory relations.

Our first task will be to determine what to take as the primary ingredients making up the logical reconstruction of a scientific theory. Since we have agreed to characterise intertheory relations both syntactically and semantically, evidently our description of a theory T will have to include its class of *models*; denote this class by M. Elements of M will be structures for some fixed, possibly many-sorted, similarity type, say τ. In addition, for the purposes of theory testing, prediction, and the like, it may also be useful to consider a broader class of structures N (of the same type) which includes M. Members of N will be referred to as *admitted structures* for T. Whilst the class M can be assumed to represent what are usually called the *laws* of a theory, there may be certain other regularities associated with T which are best construed as relations *between* models. Sneed's notion of *constraint* is one such. Another, especially prominent in physics, is the concept of *symmetry* or invariance principle. In its simplest form, a symmetry of T can be thought of as a binary relation $R \subseteq N \times N$, such that, in particular, if $R(\mathfrak{M}, \mathfrak{N})$ and $\mathfrak{M} \in M$, then $\mathfrak{N} \in M$. This indicates, in short, that the laws of T are invariant under the given symmetry.

Generalising slightly from the above considerations, one is led to picture a theory as a structure[13]

$$T = \langle \tau, N, M, R \rangle$$

satisfying the following conditions:

(1)   (i) τ is a (many-sorted) similarity type;

         (ii) N is a class of models of type τ;

         (iii) $M \subseteq N$;

         (iv) R is a class of relations of the form $\mathscr{R} = \{\langle \mathfrak{M}_1, \ldots, \mathfrak{M}_n; h \rangle, \ldots\}$, where $n > 0$, $h \geqslant 0$, $\mathfrak{M}_1, \ldots, \mathfrak{M}_n \in N$, $h = \langle h_1, \ldots, h_k \rangle$, and the h_i are relations between the individuals of the models;

         (v) N, M and R are set-theoretically definable, and N, M are closed under suitable morphisms.[14]

The conception of 'theory' framed in (1) is by no means comprehensive. However, it is, I believe, adequate as a basic unit for metascientific studies, and, in particular, for the purpose of giving an exact reconstruction of intertheoretic relations. Plainly, there is one important feature of theories still missing from our description so far: the element of *language*. But here one can follow the procedure already adopted in Chapter 1, namely that of considering the languages $L(\tau)$, for suitable logics L, that are appropriate for speaking about the models of the theory. As before, let us keep the notion of logic very general so as not to restrict unduly the possible choices of language. And, when τ is any available type for a logic L, let $\text{Sent}_L(\tau)$, $\text{Str}_L(\tau)$ denote as usual the collections of all L-sentences and L-structures (of that type), respectively; and let $\vDash_L$ be the truth relation.

In principle one can make use of any admissible logic when constructing relations between different theories. But we are chiefly interested in those logics L that are suitable for characterising the laws of T, that is to say, those for which the class of models M is L-elementary. Let us call these logics *adequate* for T. And among the adequate logics, one will usually prefer those that are the weakest, or which have certain properties that are desirable from the metatheoretic point of view. I shall assume that theories possess adequate logics in this sense, and that a suitable choice of logic can be made on pragmatic grounds, according to the aims of the reconstruction.

Given two theories $T = \langle \tau, N, M, R \rangle$ and $T' = \langle \tau', N', M', R' \rangle$, we can now offer an exact account of what is meant by a correspondence of one to the other; clearly, this relation will be relativised to some choice of adequate logics. Thus, let L, L' be logics adequate for T, T' respectively. Then,

(2)   A *correspondence of T to T', relative to $\langle L, L' \rangle$* is a pair $\langle F, I \rangle$ of mappings, such that:
  (i) $F\colon K' \xrightarrow{\text{onto}} K$, where K, K' are nonempty subclasses of M, M' respectively;
  (ii) $I\colon \text{Sent}_L(\tau) \to \text{Sent}_{L'}(\tau')$;
  (iii) K is definable in L and K' in L';
  (iv)   for all $\mathfrak{M} \in K'$ and $\varphi \in \text{Sent}_L(\tau)$,
         $F(\mathfrak{M}) \vDash_L \varphi \Leftrightarrow \mathfrak{M} \vDash_{L'} I(\varphi)$.

This definition reproduces the idea that a correspondence is composed of a structural correlation F and a translation I which respects F (clause (iv)). The logics L and L' provide the appropriate languages which the translation connects, and I include an additional assumption (ii) that these logics are adequate to define respectively the range K and the domain K' of F.

In the case of some reductive relations, one may also take a special interest in the links established between the components N, N' and R, R';[15] but I shall not give an explicit treatment of them here. Notice, however, that (2) already takes one straight to the heart of the meaning of reduction. For, it ensures that a correspondence of T to T' determines a certain continuity between these theories and even allows for the potential explanation of one of them by the other. If one thinks of K' as an 'extension' of T', formed by conjoining additional hypotheses to the laws of T', then it is clear that this extension has all the laws of T (i.e. L-sentences valid in M), under translation, among its L'-consequences.

In the event that L and L' happen to be the same logic, one can refer to the correspondence as being *relative to L*. In that case, one might consider, as a viable alternative to (2), correspondences where F itself is an L-definable operation (i.e. L-elementary or L-projective) in the sense of Chapter 1; clauses (i) and (iii) of (2) could be adjusted accordingly, and the relevant types, τ and τ', assumed disjoint (by an appropriate name-change, if necessary).

In correspondences of either kind, the correlation between the models of T and T' is, of course, only required to be partial. As was hinted earlier, some of the principal types of reduction in science can be recovered by means of natural strengthenings of this condition. To make this idea precise one can define the following relations, where T and T' are as above and $T \alpha T'$ signifies that there is a correspondence of T to T' satisfying (2).

(3) T is said to be *interpretable in T'* [*embeddable in T'*], [in *limiting case correspondence with T'*], in symbols $T \delta T'$ [$T \rho T'$], [$T \gamma T'$], if $T \alpha T'$ and $K' = M'$ [$K = M$], [for all $\mathfrak{M} \in K'$, $F(\mathfrak{M})$ is a standard approximation of $\mathfrak{M}$].

When a correspondence is both an interpretation and embedding of T in T', I call it a *faithful interpretation* and write $T \beta T'$.

The relation δ generalises the usual concept of interpretation, and can be construed as a formalisation of the notion of theoretical reduction discussed earlier. In particular, if $\langle F, I \rangle$ is an interpretation of T in T' relative to $\langle L, L' \rangle$, then it is easily seen that for all $\varphi \in \mathrm{Sent}_L(\tau)$,

$$T \vDash_L \varphi \Rightarrow T' \vDash_{L'} I(\varphi),$$

where '$T \vDash_L \varphi$' stands for '$\forall \, \mathfrak{M} \in M, \mathfrak{M} \vDash_L \varphi$', and analogously for '$T' \vDash_{L'} I(\varphi)$'. This implies, therefore, that any law (in the syntactical sense) of T has a translation that is valid in T'. On the other hand, the converse of this property plainly holds for embeddings, namely that for all $\varphi \in \mathrm{Sent}_L(\tau)$,

$$T' \vDash_{L'} I(\varphi) \Rightarrow T \vDash_L \varphi.$$

And this indicates that ρ can be regarded as a formal counterpart of explanatory reduction. To spell out the situation in somewhat more detail, consider the potential explanation by T of a 'fact' or 'problem' expressed by an $L(\tau)$-sentence φ. Let the explanans from which φ is to be derived be represented by the conjunction of 'laws' $\theta_1, \ldots, \theta_n$ and 'initial conditions' $\psi_1, \ldots, \psi_k$. Since the explanans, to be consistent, must be satisfied by some model $\mathfrak{M}$ of T, it follows that the translations $I(\theta_1), \ldots, I(\theta_n), I(\psi_1), \ldots, I(\psi_k)$ are jointly satisfied in a model $\mathfrak{M}'$ of T', where $F(\mathfrak{M}') = \mathfrak{M}$. The conjunction of these translated sentences is thus consistent with T' and forms the basis for an explanation by T' of the (translated) explanandum $I(\varphi)$.[16]

The γ-relation embodies a different type of constraint on the structural mapping F. It can be viewed as a formalisation of the core of approximative reductions, where the notion of *standard approximation* provides a formally precise characterisation of the intuitive idea that the correlated models are 'close'. I shall not go into further details for the present, but later, in Chapter 7, an application of limiting case correspondence will be discussed.[17]

The reductive relations defined so far place no special restrictions on the vocabularies of T and T'. Embeddings and interpretations, for example, may hold between theories that possess quite different sets of primitive terms, as well as between theories that share a good deal of logical and nonlogical vocabulary. In the event that one of the theories contains *all* the primitives of the other, one can distinguish important special cases of embedding and interpretation, as follows.

In general, when one speaks informally of a theory T 'extending' another theory T', this may mean that T contains additional laws, or that it has additional concepts (or both). In the former case, the class of models of T' is narrowed down; in the latter case, the similarity type of the models of T' is expanded. Moreover, T may, in the first sense, 'extend' T', even if it contains only a part of the vocabulary of T'; and T may, in the second sense, 'expand' T' in a quite inessential way; for instance, if the extra primitives of T turn out to be 'definable' in the language of T'. In our framework of reductive relations, useful versions of these last two special types of theory extension can be introduced like this[18]:

(4)   T is said to be an *extension* of T', in symbols $T \sigma T'$, if $T \rho T'$, $\tau \subseteq \tau'$, I is the identity mapping, and F is the restriction of the reduct function, $\upharpoonright$, that takes each τ'-structure to its τ-reduct.

T is said to be an *expansion* of T', in symbols $T \varepsilon T'$, if $T \rho T'$, $\tau' \subseteq \tau$, $K' = M \upharpoonright \tau'$, the restriction of I to $\text{Sent}_L(\tau')$ is the identity mapping, and for all $\mathfrak{M} \in K'$, $\mathfrak{M} = F(\mathfrak{M}) \upharpoonright \tau'$.

And the 'dual' relations, σ' and ε', can be defined from (4) by requiring for the former that $K' = M'$ and for the latter that M'

$= K \upharpoonright \tau'$, instead of the conditions $K = M$ and $K' = M \upharpoonright \tau'$, respectively.

The sense in which the newly introduced relations are genuinely reductive needs some qualification and explanation. According to the standard intuition, the reducing theory is the 'stronger' of the two in terms of laws, and the more 'comprehensive' in terms of concepts. (4), on the other hand, permits an extension T of T' to possess additional laws, and an expansion T of T' to possess additional concepts. And yet each of these relations is a special case of embedding, which we have just construed as an exemplary instance of reduction. (Analogously, σ' and ε' are special cases of the interpretation relation.)

From the logical point of view this situation is not anomalous, however. It is quite legitimate in reduction contexts to assume that auxiliary hypotheses may be required in order to effect a 'deduction' of the reduced theory from the reducing theory. In this sense, any extension (or expansion) T of T' is trivially reducible to T' when the 'additional strength' of T appears in the guise of auxiliary hypotheses; just as any theory is vacuously reducible to itself. From a broader, methodological standpoint, it is clear of course that one has to exercise caution when assigning an 'explanatory' function to the reducing theory. It would be unusual to regard a theory as 'explaining' itself or one of its extensions, in anything other than a purely gratuitous sense of the term.

In general, however, whether one can speak here of a genuine reduction of T to T' — e.g. in the sense of T' explaining T — depends on the context involved. In particular, it depends on whether acceptable auxiliary assumptions exist (not involving the full strength of T) which, when conjoined to T', would enable T to be derived. When appropriate assumptions are available, one may well speak of a genuine, nontrivial and noncircular reduction of T to T, even if T is an extension or expansion of T'.

Here is an example. One can represent Newton's gravitational theory as an expansion of the basic, core theory of classical mechanics involving Newton's three laws of motion. However, in some prominent textbook accounts (e.g. Born, 1962) the law of universal gravitation is *derived* from the laws of motion in conjunction with (certain well-confirmed consequences of) Kepler's laws. Since Kepler's laws themselves are logically incompatible with Newton's, this 'derivation' has been the subject of some controversy. Recently, though, Elie Zahar (1983) has argued forcefully that not only did Newton indeed arrive at his gravitation theory by just such a 'deductive' route, but that, providing appropriate care is taken, one may reconcile the various steps in the derivation with the usual canons of deductive reasoning. Under this interpretation it would seem perfectly respectable, therefore, to regard Newton's gravitation theory as being explained from, and reducible to, his basic force laws. And a similar view might also be defensible in the case of some other applications of classical mechanics comprising additional force laws, where the latter are depicted as expansions of the basic theory. If so, then at least for some of the principal cases of interest (especially those discussed in Chapter 5 below), a genuine reduction of the expanded theory may be obtained.

From a wider perspective, one can simply view the notions introduced above as amounting to a collection ITR of intertheory relations of a broadly 'reductive' sort. Thus, let ITR = $\{\alpha, \gamma, \rho, \delta, \beta, \sigma, \varepsilon, \sigma', \varepsilon'\}$. The members of this collection can be partially ordered according to logical strength. Among binary intertheoretic relations, we say that α_i is *stronger than* $\alpha_j (\alpha_i \geqslant \alpha_j)$ if for any theories $T, T', T \alpha_j T'$ whenever $T \alpha_i T'$. Thus the elements of ITR can be compared as in Figure 1, where the stronger relations appear above the weaker.[19]

The set ITR is by no means intended to exhaust all types of intertheoretic relations in science. It provides, however, a reasonably rich catalogue of relations that can fruitfully be applied in

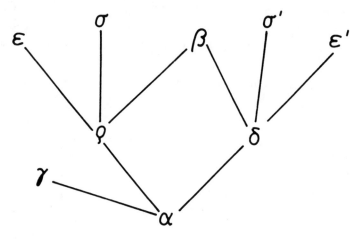

Fig. 1.

reduction studies, and it offers a classification and comparison of different reduction types. There are several directions in which ITR may be enlarged; I shall mention some of them in outline.

In the first place, one may strengthen the conditions on embeddings and interpretations so as to yield relations that are intermediate between ρ and σ, ε (and between δ and σ', ε'). An example would be to add the constraint that the translation I when restricted to the shared sentences (of $L(\tau)$, $L'(\tau')$) is (up to some equivalence) an identity mapping. Though, as I argued at length in Chapter 2, this requirement of *literal* translatability cannot be considered a necessary feature of reducibility, it may nevertheless arise naturally in some contexts. For instance, reduction relations between theories belonging to the same paradigm or research tradition might be of this type, where there are no radical changes of meaning involved.

A second sort of refinement consists in imposing specific restrictions on the ways in which the 'higher-order' relations R and R' are to be connected. The case of Sneedian *constraints* is dealt with at length in the structuralist literature; the case of *symmetries* has received an explicit treatment by Rantala and myself, elsewhere.[20]

Thirdly, one might consider additional binary (and more generally *n*-ary) intertheory relations, formed by taking suitable compounds of the existing relations. For example, define $T \; \alpha\varepsilon \; T'$ by $\exists T_1, \; T \; \alpha \; T_1 \; \varepsilon \; T'$, to be read as '$T$ is in correspondence with some expansion of T''; and other hybrid relations can be characterised analogously. This illustrates some of the ways in which ITR can be transformed into a comprehensive collection of relations. In some respects, however, the classification of basic relations in ITR still remains coarse, and I shall discuss further possible refinements below.

4. FURTHER PROPERTIES

When introducing a new explication, or a systematic analysis, of a familiar concept in the philosophy of science, there are of course important methodological questions to confront. How does the new explication differ from and, in particular, improve upon previous accounts? Where do its chief strengths and weaknesses lie? Does it satisfy reasonable criteria of adequacy? Does it contribute to the clarification of existing philosophical problems, and perhaps open up interesting new lines of research? Does it assist us in the rational reconstruction of science and its development? Naturally, the network of intertheory relations proposed here is bound to be subject to questions of this kind. In the remainder of this chapter, and to some extent also in the next, I shall offer preliminary responses to some of them.

Taking a broadly conceived typology of reduction as our starting point, we have so far settled on a formal framework of intertheoretic relations each of which is based on, or analysable in terms of, a core relation of *correspondence*. In the first place, it is plain that correspondence represents a generalisation (in various respects) of the dominant accounts of reduction. Compared to the classical, syntactical-derivational schema of reducibility, we have extended the ideas of translation and deduction to make explicit how logics of different kinds and strengths may be involved in relating two or more scientific theories. We have also extended the structural approaches of the Suppes and Sneed variety, by requiring a reduction functor to be not merely set-theoretically definable, but (in essence) model-theoretically definable too; once again, the relevant logics being indicated. And, in another sense, we have refined and reinforced both of these two accounts by feeding a semantic component into the former and a syntactic component into the latter.

Correspondence thus comprises both a semantic correlation F and a syntactic translation I. Imposing further conditions on F and I led, as we saw, to some standard types of reducibility. Above all, aspects of theoretical and explanatory reduction were highlighted in this manner, through the relations of interpretation and embedding. It is easy to see, however, that information regarding other properties of F and I can be exploited in analysing conceptual and ontological forms of reduction. Rather than incorporating such properties directly into our taxonomy of relations, it seems sufficient (at the present stage of investigation) to be aware of them, and to study their effects (when they are present) in concrete cases.

Consider first the structural correlation F. In some cases F may associate elements (from the domains) of a model $\mathfrak{M}$ with the *same* elements of $F(\mathfrak{M})$. Or it may associate elements of $\mathfrak{M}$ and $F(\mathfrak{M})$ in a one—one fashion; or determine in a more complex manner a construction of $F(\mathfrak{M})$ objects out of the entities in $\mathfrak{M}$. The questions whether such correlations between objects represent 'identities' or some other kind of linkage, whether they hold necessarily or contingently, whether they represent lawlike or accidental generalisations, and so forth, do not, properly speaking, belong to the logic of reducibility. Yet it is clear that the logical form of the link which F establishes between objects in the domains of T'-models and those in T-models is relevant to the issue what kind of ontological reduction, if any, is determined. In other words, the nature of potential ontological reductions of T to T' can be precisely characterised in terms of the logical properties of the map F. Examples of appropriate questions to ask here would be: Is there a concrete construction taking $\mathfrak{M}$ to $F(\mathfrak{M})$? Are the elements of $F(\mathfrak{M})$ defined from elements of $\mathfrak{M}$, and, if so, by what type of definition?

What of conceptual reduction? It is customary here to think in terms of connecting hypotheses, often called bridge laws. On the surface, bridge laws and the like are no longer present when one switches to the more general representation involving the translation map I. But, evidently bridge laws can, if desired, be reintroduced; for instance they can be treated as a subclass of expressions ψ of the form

$$\psi := (\varphi \leftrightarrow I(\varphi)),$$

for suitable sentences φ in the language of T. More generally, if I is extendible to a mapping defined on open formulas, a bridge law would typically be represented by the universal closure of ψ, for certain atomic formulas φ,

One of the standard philosophical issues associated with reduction has been the epistemological status of bridge laws: Are they empirical principles, or perhaps conventional truths? For the purposes of grasping the nature of conceptual reduction, however, such questions look decidedly sterile. Our present perspective invites us to consider a more revealing set of questions. There is, for example, no *prima facie* guarantee that expressions like ψ above will be sentences of any well-defined language, much less that such a language would enjoy a definite semantic interpretation. On the other hand, one is always at liberty to investigate the logical form of the translation I, and thereby determine what kind of reduction of concepts it embodies. Rather than discuss the 'status' of the claim, say, that temperature can be equated with mean kinetic energy, it seems more fruitful, therefore, to study the possible methods of translating the language of thermodynamics into the language of statistical mechanics, and to look at the structure of the corresponding microscopic explanations of thermodynamical laws and phenomena which such translations give rise to.

To sum up: the philosophical literature on reduction abounds with debates over obscure metaphysical questions of 'identity' and 'necessity' that are held to arise when the concepts and ontologies of different theories and frameworks are related. Lacking clearly formulated problems and adequate instruments for solving them, these debates have brought about scant consensus and achieved little in the way of concrete results. By contrast, the principal problems surrounding the different forms of conceptual and ontological reduction and the structure of reductive explanations have, until quite recently, received less than their fair share of attention, probably due to the absence (or ignorance) of appropriate logical methods for tackling them. My contention is that the general model-theoretic framework sketched here prompts the right sort of questions to pose in reduction contexts and offers a suitable setting for their solution. In characterising the set ITR, I have, of course, employed only broad distinctions related to the syntactic and semantic functions, F and I. But it is clear that, by examining in specific cases the fine structure of these two mappings, one can read off a wealth of additional information regarding the nature of the reductions involved.

5. CRITERIA OF ADEQUACY:
SOME FALLACIES EXPOSED

The last point about individual variations displayed by different translations and structural correlations is closely connected to another theme raised earlier: the problem of *adequacy*. It is common currency that an adequate explication of reduction ought to set standards whereby all 'genuine' instances of reducibility are admitted, and any fortuitous instances excluded. This desideratum looks reasonable at first sight, but in fact a number of questionable assumptions underlies it. Among them is the implicit supposition that there is a single explicandum, 'reduction', to be analysed. To claim, as I have, that different types of reducibility share some common properties, is not necessarily to grant the uniqueness of the explicandum. Furthermore, even if the label 'reduction' stood for a single concept, one should not assume it to be unambiguous. Indeed, it would be a fallacy to suppose that the informal notion of reducibility possesses a well-defined and well-understood extension comprising exactly the class of theory or law pairs with one member of each pair standing in the said relation to the other. On the contrary, the principal aim of the reconstruction is to replace an intuitive but vague concept by a more exact set of concepts that allows for an accurate and informative description of the relations holding between actual theories. In this sense, the explication might not only refine but also reform existing usage: it might provide grounds to reclassify or even rule out some previously accepted examples of reduction, as well as to admit new examples into the fold.

The criterion of adequacy just cited cannot, therefore, be interpreted literally; and this applies *inter alia* to analogous criteria invoked in connection with other systematic explications in metascience which aim to be critical as well as descriptive. At the other extreme, however, there is a danger that an overly radical departure from standard usage might leave one open to the charge of explicating a notion *different from* the intended one or ones. How then can we balance the need to accommodate well-established cases of reduction in science with the desire to exclude the trivial or gratuitous cases?

This issue has lately been revived as a result of attempts to criticise some of the traditional accounts of reducibility. For example, Hoering (1984) argues that both the syntactic and the semantic approaches to reduction can be shown, on purely formal grounds, to admit counter-intuitive instances. This, he claims, applies in particular to the idea of reduction as relative inter-pretability (in the Tarskian sense) and as a functional correlation between models (along Sneedian lines).[21] Even if correct, this criticism would not apply without further argument to our schema ITR of intertheoretic relations, since we rely on a combination of syntactic and semantic conditions for reducibility, and thereby run less risk of admitting unwarranted cases.

All the relations in ITR are thus of a reductive sort, since, in one way or another, the various logical and epistemological types of reduction can be subsumed under them. As we have seen, even the basic relation of correspondence may characterise a type of intertheory explanation. But it might be objected nonetheless that the syntactic and semantic conditions required for correspon-dences, interpretations, embeddings and the like, provide at best *necessary* conditions for genuine reductions between scientific theories, but that they cannot by any means be considered *sufficient*. Our analysis thus far would thereby fall short of a proper explication of reduction.

This type of objection has sometimes been offered up as a general criticism of 'purely formal' explications of reduction; Hoering's paper being only one recent example of this sentiment. At least in its most common form, however, the objection seems to me to be quite unfounded. It is, I believe, based on a misunderstanding of formal approaches to reduction of the sort sketched here, and, in part, on a questionable view of what it means in general to explicate a given metascientific concept.

The argument I wish to rebut usually takes the following form. The conjunction of suitable syntactic and semantic constraints (e.g. of the kind given here) will never yield sufficient conditions for reducibility in science because there will always remain additional and 'irreducible' nonformal criteria that successful reductions should meet. In particular, this means that the formal conditions could be gratuitously satisfied in cases where we would never speak of a 'genuine' reduction (given that certain other criteria are not fulfilled). Applied to ITR, this would entail that establishing the presence of, say, an embedding or an interpretation between theories cannot be a 'test' of reducibility.

Since concrete examples of such allegedly fortuitous cases of reduction are hard to come by, this argument seems to rest for its plausibility on the strength of the claim that certain ingredients of reduction cannot be 'formalised'. My own view, however, is that the very distinction being drawn here between formal and nonformal conditions for reduction is untenable and rests on a fallacy. Let us consider in turn three special types of nonformal requirements.

The first case is one already touched on earlier: a theoretical reduction is supposed to supply an explanation of the reduced law or theory. In this sense it seems that the reducing theory, together with any further assumptions belonging to the explanans, should satisfy certain epistemic requirements of acceptability. This constraint, it is claimed, is not reflected in any structural property of the reducing theory and must, therefore, be added as a further, nonformal criterion for reducibility.

In reply, I would agree that (under the stated assumptions) the above requirement is a reasonable one and may, as Nagel (1961) has argued, be an important feature of reductions of the 'explanatory' kind. Rather than strictly 'nonformal', however, the requirement seems to be *pragmatic*, and, as such, I see no intrinsic barrier to its formalisability. Perhaps even the more far-reaching pragmatic approaches to scientific explanation, of the Gärdenfors (1980) kind, may also find application in reduction studies; but it seems to me that the syntactic/semantic aspects of reduction are liable to remain the dominant ones. As remarked earlier, to speak of T' explaining T in the case of a simple correspondence, $T \alpha T'$, we may need at least the extra premise that any auxiliary assumptions involved are well-grounded. But it should be noted that in cases where no independent empirical 'test' of such assumptions is available, the success of the correspondence itself may supply confirmatory evidence. When, on the other hand, T is interpretable in T', no ancillary hypotheses are required for T' to act as an explanans and we can rest content with the epistemic acceptability of T' alone. Here too, however, it may happen that T' is not an independently testable theory, but that evidence for it may accrue indirectly from its capacity to entail one or more well-confirmed laws or theories. In this context, Nagel (1961) cites the kinetic theory of gases as an example of a theory whose chief support derives from its reductive credentials, in particular from the fact that the Boyle—Charles law can be deduced from it.[22]

A second and closely-related issue is also raised by Nagel under the rubric "nonformal conditions for reduction". A correspondence relation will seldom carry decisive weight for establishing a genuine reduction unless the relation can be seen to apply to a variety of different cases (laws, theories). In our terminology, the point might be restated this way: A translation (and structural correspondence) between given scientific languages will gain epistemic support when, by means of it, reductive relations between different pairs of theories (or from different theories to a given one) can be set up.

This is also an important feature of reduction, with a clearly pragmatic component. It cannot be formally embodied in our reconstruction of ITR as long as we deal only with single laws or 'small' theories. But once we treat comprehensive theories, like classical and relativistic mechanics, or even entire research traditions, the matter is quite different. In fact, one of the advantages of the present approach, as compared with, say, the structuralist one, is that by allotting translation an explicit role in reduction, it is a simple matter to formulate further conditions requiring that a given translation be repeatedly employed to establish reductive links between different laws and 'theory-parts'. I shall return to this problem in the next chapter, where research traditions will be treated explicitly.

A third type of nonformal constraint I want to consider is less easy to state clearly. Some authors express it very loosely as the requirement that reduction should connect only theories that are in some intuitive, or perhaps 'pre-theoretic', sense "about the same things". Unfortunately, it seems impossible to turn this into a precise and independent condition for reducibility. Suppose, for instance, we were to demand for $T \alpha T'$ that T and T' should share the same 'problem-domain' or the same ontology, understood in some pre-theoretic sense. This would exclude at a stroke many genuine and scientifically fruitful cases of reduction; all those cases, in fact, where *prima facie* different domains and ontologies have been unified through reduction, or where one has been subsumed under another. Such semantic or ontological 'similarities' cannot, therefore, be pre-theoretic conditions for reduction: they are, if at all, features which the reduction itself must reveal.

Hoering (1984) has attempted to express one aspect of the above idea in somewhat sharper form. Besides considering only the *representations* of physical, biological and other types of systems by means of semantic structures or theory models, we should, he argues, consider also the *real* systems themselves. His chief proposal for a nonformal condition for reduction then amounts in our notation to roughly this: two structures $\mathfrak{M}$ and

$F(\mathfrak{M})$ correlated by means of a reductive correspondence F should be representations of the *same* (physical, biological, etc.) system.

> ... more exactly: every description in the language of the reduced theory of a set of systems becomes a description of the same set of systems, when translated into the language of the reducing theory.
> This is not a condition which can be rendered in syntactical terms.
> (Hoering, 1984, pp. 45—46)

In reply to this proposal, one should first note that it is not always evident that this condition can be clearly understood; in particular, it is not obvious that one can speak meaningfully about the identity of 'real systems' independent of their description or representation within some theory. Suppose, for example, we had good grounds to believe in psychophysical reductionism. Would this license us to claim that every mental state is determined by a *corresponding* physical state? Or would it support the apparently stonger claim that every state describable in psychological terms is fully determined by *its* physical characteristics (and is thus identical with a physical state)? Would there, in fact, be any meaningful difference at all between these two formulations?[23]

If there is sometimes a problem in giving an independent identification of 'real systems', at the same time there is a perfectly legitimate and mundane sense in which reduction may concern itself with the same systems. Consider the reduction of the gas laws to the kinetic theory, or that of the mechanics of rigid bodies to the mechanics of point particles. If these reductions are successful, we should certainly expect (with appropriate qualifications) that any fixed experimental situation — a configuration of rigid bodies, say, or a given volume of gas in a closed container — should be uniformly the subject of description by both reduced and reducing theories. This means, in effect, that if the situation in question is represented in one theory by a certain model $\mathfrak{M}$, that same situation will indeed be represented in the other theory by its counterpart model, $F(\mathfrak{M})$.

One can agree with Hoering that the above condition is not syntactically expressible within the languages of the theories concerned. It is, rather, a metatheoretic constraint. But, as such, it can and usually does form part of our semantic requirements for correspondence, interpretation, etc. The point is that, in one sense or another, we require the structural correspondence F to be *definable* (as well as accompanied by a respectful translation). To win maximum flexibility, we have been generous in our construal of definability; nevertheless, F is intended to express a canonical relation between the reduced and reducing models. Thus, where concrete applications are concerned, it is to be expected that $F(\mathfrak{M})$ is definable or constructible from $\mathfrak{M}$; and this means, in particular, that the domain of $F(\mathfrak{M})$ is definable or constructible from that of $\mathfrak{M}$. When, for instance, rigid bodies are correlated with point particles in the reduction just cited, those particles represented in a model $\mathfrak{M}$ are the stuff of which (up to some equivalence) the rigid bodies in a corresponding model $F(\mathfrak{M})$ are 'composed'. If the reduction is correct, there is no question, therefore, of an experimental configuration described within the reduced theory not being associated by the correspondence with the 'same' configuration now differently described within the reducing theory.

That we do not lay down in advance the exact nature of the definitions or constructions involved is dictated by the desire to be 'open-minded' about admitting reductions that might be scientifically respectable, if not now, at least in the future. As we saw in Chapter 2, there are promising candidates for characterising a 'reasonable' class of constructions: for instance, the class of Gaifman operations, or the class of algebraic constructions admitted by the 'elementary interpretations' of Sette and Szczerba (1978; 1983). The latter involve finite iterations of definitional extensions, restrictions, Cartesian products and quotients. But for the reasons already given, the restriction to interpretations that are, say, first-order definable, or parameter-free, seems less attractive once we move outside the field of elementary algebra to treat reductions in science at large.

The history of science shows plainly enough that almost no cognitive domain of investigation can be considered beyond the reductionist's clutches. Whilst we should to try to set 'reasonable standards' for the reductionist to meet, it is therefore virtually impossible to exclude *a priori* potentially fruitful or successful reductions on purely intuitive grounds of 'plausibility' or 'similarity'. This applies even to radical reductionist theses like the one sometimes associated with Artificial Intelligence, and discussed, for example, by John Searle (1984) under the label "the strong AI thesis". It amounts roughly to the claim that cognitive processes in the human brain are instantiations of a computer program. However *prima facie* implausible the thesis might be, it can ultimately be quashed only by arguing that AI does not yield adequate explanations of human mental processes[24]; it cannot be put to rest merely by appealing to the obvious physical differences between human brains and computers.

In short, then, proposals of Hoering's sort are already contained in the spirit, if not in the letter, of our characterisation of reductive relations, ITR. If desired, they can readily be incorporated explicitly, not as intrinsically nonformal conditions for reducibility, but as perfectly ordinary semantic constraints. As it stands, however, our explication of reduction is open to a 'nonconstructive' as well as a 'constructive' or 'effective' reading. Perhaps, in some cases, an argument for interpretability or embeddability could be based on some nonconstructive existence proof, rather than an effective presentation of mappings F and I having the requisite properties. This is no real disadvantage, however, providing that the explication is properly used; providing, that is, one is careful not to assert more than what has actually been demonstrated.

By way of illustration, consider, once again, psychophysical reductionism as a potential example. It is not inconceivable that one day a proof by noneffective means could be given that an acceptable theory of cognitive psychology is interpretable in a suitable neurophysiological theory which is, in its turn, interpretable in some well-confirmed portion of physics. This kind of result might not satisfy a fully-fledged materialist or psychophysical reductionist (or for that matter his critics, though reductionism has often been argued on far slenderer grounds than these), but it might provide a valuable elucidation of some aspects of the mind—body problem. It would, in particular, show that psychological laws are at least in principle derivable from the laws of physics, even if there were no way of knowing *how* to derive them. The result might therefore be used to support some versions of materialism, and, more importantly, it might be heuristically fruitful in suggesting directions in which genuine physical explanations of mental phenomena should be sought.

The situation is broadly similar with respect to the pragmatic features of reduction discussed earlier: these too may be invoked to support or to question alleged instances of reduction. Nevertheless, we can defend the present characterisation of ITR and claim that in each case necessary and sufficient conditions are supplied for the designated relations. We can also claim, I believe, that this explicates (different senses of) the word 'reduction'. It is essential, however, to keep two aspects of this reconstruction firmly in mind. First, reduction is not taken to be a single explicandum, and is therefore best explicated by means of a family of intertheoretic relations. Secondly, the present reconstruction does not provide an exhaustive list of all the features that might be relevant in analysing alleged cases of reduction; sometimes finer distinctions have to be made in order to resolve disputes about reducibility. But the analysis does suggest several kinds of syntactic, semantic or pragmatic factors (e.g. the recursiveness of a translation, the effectiveness of a model-theoretic construction, the range of applicability of a translation, etc.) which may indeed be relevant in a given situation.[25]

In general, the system ITR can best be conceived as providing a framework and a set of guidelines on matters of reducibility. It distinguishes certain common types of reduction, and allows one to study different degrees of reducibility. As with any metascientific reconstruction, one is not free to step too far out of line with ordinary usage; though, where appropriate, one can certainly endeavour to sharpen and even to reform standard practice. Since intertheory relations are not only tools of rational reconstruction, but can also be interpreted as rules and determinants of scientific research, one's characterisation of them should be sufficiently flexible to accommodate fresh insights and changing intuitions gathered from the actual development of the sciences.[26]

5

THEORY DYNAMICS, CONTINUITY
AND PROBLEM-SOLVING

1. INTRODUCTION

Contemporary epistemology and philosophy of science devote increasing attention to the following types of question. Do the various sciences progress continuously, or in discrete stages? Does scientific knowledge grow cumulatively, or dispersively? Can scientific change best be described by means of an evolutionary model, or does the Kuhnian pattern of normal science, crisis and revolution constitute a superior fit?

These and similar questions about scientific development are all, of course, variations on a single theme. Without further qualifications they are also equally intractable, indeed unanswerable. It seems obvious that against a stable background framework of rational reconstruction, some episodes of scientific development will seem continuous, others may not. Conversely, a particular case of theory change in science may appear rationally well-founded and even cumulative from one metascientific perspective, whilst a quite different picture may emerge when our metatheoretical stance is shifted. Moreover, 'continuity' is by nature bound to be a relative concept, a matter of *degree*. At best we might hope to indicate that the transition from one theory to another is *more or less* smooth, or, at the other extreme, *more or less* abrupt.

One of the pressing tasks of the philosophy of science is thus to impose some order onto this conglomerate of problems by establishing appropriate parameters by which constancy and change can be assessed. Those philosophers with a leaning towards formal methodology may prefer to work with logical categories and technical instruments. Others may look towards history, psychology and sociology of science to provide the tools for understanding, explaining and evaluating scientific progress. It is important to realise, however, that the logical and the historical approaches to the rational reconstruction of scientific development possess reciprocal rather than conflicting aims and methods. Each may, and from time to time does, learn from the other. The introduction of a logical formalism may be a valuable refinement for a primarily historical model of theory evolution. By the same token, the addition of pragmatic and historical features to a formal model of theory structure and theory dynamics may show it to be sensitive to the lessons learnt from studies in, say, the history or the sociology of science.

We saw earlier how Stegmüller, Balzer, Moulines and others working within the structuralist programme have taken considerable pains to formalise certain aspects of Kuhn's (and also Lakatos') account of the development of science. To the extent that this enterprise is successful, it may be beneficial to both parties. For structuralists, it may show how their framework is no empty formalism, but an instrument that can be profitably applied to the historical dimension of scientific inquiry. From the other side, it may indicate to historians and ordinary-language philosophers how some of the central methodological concepts can admit a precise and rigorous characterisation. In a fruitful dialogue between the two types of approach, historical research will suggest the directions in which a formal framework can usefully be extended and improved, whilst logical reconstructions may serve to test and sharpen concepts and ideas acquired from historical analysis.

In earlier chapters I expressed some strong doubts and reservations about the adequacy of the structuralist framework. I focused there largely on problems of intertheory relations and the inability of the structuralist approach to deal correctly with matters of translatability and commensurability. It can readily be seen, however, that these difficulties are further aggravated when the structuralist framework is applied to science as a diachronic process. Taken as a whole, therefore, Stegmüller's programme of 'theory dynamics' cannot be viewed as a complete success.[1] But, though there is reason to be sceptical about many of the details of this programme, it does contain certain basic ideas and insights that are worthy of further pursuit. Among them is the notion, present already in Sneed's (1971) treatment, that often a comprehensive physical theory or paradigm can be perspicuously represented as a structured family of interconnected theories or 'theory-elements'. This permits, among other things, a clearer characterisation of the cognitive content of a scientific paradigm and its component theories.

I shall not further discuss here the merits and shortfalls of the structuralist approach to these issues. Instead, I want to consider some of the ways in which the general model-theoretic framework of the last chapter can be extended and applied to deal with aspects of problem solving within a scientific research tradition. The aim of this exercise is not so much to present a formalisation of Laudan's theory, but rather to show how certain types of continuity or discontinuity in scientific change can be given a fairly precise description. To this end, we shall make use of some core features of Laudan's model of growth, but at the same time try to fill in some of the gaps and correct some of the inadequacies of that account which emerged from the discussion in Chapter 3.

In what follows, then, we shall embark on a preliminary investigation of two, closely related, issues: How can we achieve a more exact representation of problem-solving within a research tradition? And, how can we establish that a problem formulated and solved within one tradition is also expressible, and either solved or anomalous, in another tradition? Both questions seem to require a more careful and detailed response than Laudan himself provides. As to the first, it is of course correct but uninformative to say that a research tradition solves a given problem when that problem has been adequately answered by some theory in the tradition. The important point is that a research tradition will impose certain constraints on what is to count as an acceptable theory and a successful problem solution. Only by making these constraints more or less explicit can we assess the long-term capacity of the tradition to evolve progressively and increase its problem-solving effectiveness.

The second question becomes important as soon as we try to compare the relative progressiveness of two or more rival research traditions. If we concede, as we should, that rival traditions may employ quite dissimilar conceptual frameworks, not only their solutions to common problems but even their modes of expressing such problems may be very different. Consequently, identifying a problem as *a shared problem* is a nontrivial task and an essential prerequisite for an eventual analysis of comparative progressiveness.

The model-theoretic framework of the last chapter can, I think, be exploited to shed light on both these questions. The method consists in trying to give an account of the basic, inner structure of a research tradition, and then to define some uniform logical relations between traditions that will allow for an explicit treatment of problem-sharing.[2]

2. ASPECTS OF PROBLEM-SOLVING

Recall that in Laudan's model of scientific progress, the solved and anomalous empirical problems, together with conceptual problems, form the basic cognitive index for rationally appraising theories and research traditions. In so doing, the *shared* empirical problems provide an essential cognitive link between rival theories and traditions, and yield a measure of the continuity of scientific change. As we already noted, Laudan leaves 'the problem' as an undefined primitive term; but he is somewhat clearer about the nature of the problem-solving *process*: "In the simplest cases, a theory solves an *empirical* problem when it entails, along with appropriate initial and boundary conditions, a statement of the problem" (Laudan, 1981a).

In other words, a problem p is solved by a theory T if, given suitable auxiliary assumptions σ,

(1) $T, \sigma \vDash \varphi,$

where φ is a sentence expressing p.

In several respects, schema (1) is an obvious simplification. We usually tend to think of problem-solving as a question-answering process, whereas in (1) the sentence expressing the problem already represents, so to say, the *answer*. A more complete picture would therefore be obtained by including also a specific question q to which φ would be the response. This type of refinement is present, for example, in Jaakko Hintikka's *interrogative* model of scientific inquiry.[3] Hintikka suggests, moreover, that the logic of such a question-answering process differs in some respects from the standard deductive logic of scientific explanation. Naturally, once questions are explicitly introduced in this way, they can be classified according to type (e.g. why-questions can be distinguished from how-questions), and the interrogative process can be analysed with the help of well-established formal methods, taken, for instance, from the field of erotetic logic.

For present purposes we can nevertheless retain (1) as a first approximation to the problem-solving process; since, though simplified, it seems to capture the essential structure of an important class of scientific problem solutions, namely those concerned with *explanation-seeking* questions.[4] A more complete reconstruction of the problem-solving model of scientific progress would evidently deal in greater depth with the question-theoretic aspects.

Where appropriate (1) can also be generalised in different ways. For instance, we might wish to represent the additional premises not as a single condition σ, but as a conjunction of distinct hypotheses possibly playing different roles in the solution. We could distinguish here between, say, initial assumptions relevant to the particular problem situation, additional general laws valid for the domain in question, limiting and boundary conditions, and so on. We might also add further information to qualify the sense in which the solution φ is to be inferred. Does it suffice that φ be a logical consequence (in some logic) of T and σ? Does an explicit argument or proof of φ need to be exhibited; and, if so, is the complexity of the proof relevant in assessing the success of T in solving the given problem? Again I shall assume that these kinds of additional factors can, where necessary, be included at a later stage when reconstructing the problem-solving model.

For a given problem p to be instrumental in determining the respective performances of rival theories or research traditions, it must be possible to establish whether p is expressible in, solved by, or anomalous for either theory. Clearly, a problem may be formulated in different ways and expressed within different languages and conceptual schemes. We need to know, then, in what sense two distinct sttaements φ and φ' might be said to express the same problem p, where φ and φ' are sentences from possibly distinct languages and theories. And, conversely, a single statement φ might express two different problems, if it is ambiguous which framework or theory is to provide the intended interpretation of φ.

Laudan, as we already remarked, gives no advice for how to

relocate an old problem in the context of a new theory. Suppose that T and T' are theories from rival research traditions RT and RT' respectively, and that a problem p is formulated by a sentence φ in the framework of T and by a sentence φ' in the framework of T'. Laudan merely suggests that, at least in the majority of cases, it will be possible to identify φ and φ' as expressing the same problem, independently of the two research traditions in question. (This leads, in particular, to an analysis of comparative problem-solving in the manner depicted in Figure 2.)

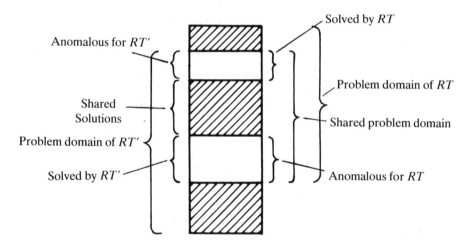

Fig. 2. Laudan's analysis of research tradition interaction.

As I argued in Chapter 3, Laudan's assumption here is a dubious one that can and ought to be relaxed. A more plausible hypothesis would be that φ and φ' are correlated by means of a translation defined between the languages of T and T' or of RT and RT'. It may well be sufficient that the translation is a partial one, not necessarily defined on all sentences of the theories involved. And it may happen that such a translation is in a certain sense independent of the specific claims of T and T' without

being independent of all the theories in the two traditions. Later
on, in Chapter 7, I shall discuss partial translations of this type in
greater detial. For the present, however, let us assume that the
theories T and T' are connected by one of the intertheoretic
relations in ITR defined in the last chapter. In other words, let us
suppose that there is at least a correspondence $\langle F, I \rangle$ between T
and T', and hence, besides a structural correlation, also a
complete translation of all sentences in the language of one of the
theories (say T) into the language of the other.

Under this construal, then, let p be a problem expressed in T
by the sentence φ, and solved under the schema

$$\theta, \sigma \vDash_L \varphi,$$

where, for simplicity, we take T to be axiomatised by a single
sentence θ, where σ axiomatises the additional premises or
boundary conditions, and where L is a logic adequate for T.
Suppose that T' has L' as an adequate logic and that $\langle F, I \rangle$ is a
correspondence of T to T' relative to $\langle L, L' \rangle$. Then a corresponding solution of p by T' might be represented by

$$(2) \qquad \theta', \sigma' \vDash_L I(\varphi),$$

for a suitable choice of conditions σ'. When (2) holds we might
be inclined to say that T' has successfully solved the problem p
transferred to the new framework.

In order to delve deeper into the role of problem-solving in
theory change, we should, however, turn to the broader context of
the research tradition as a whole. To this end we need to characterise not just single theories, but whole families of interrelated
theories.

3. RESEARCH TRADITIONS AND THEORY ENSEMBLES

Modern studies in the history of science have had a profound impact on the way in which philosophers and methodologists of science currently think about 'progress'. Today it is usual to regard progress as a comparative notion. That is to say, given some pool of competing theories, our primary task is to determine which of them is the *more* progressive. At the same time, progress has to be related to a wider *context* such as a research tradition, paradigm, or whatever; so that the empirical successes and failures of a theory are not the only factors determining progressiveness. And, thirdly, for a rational appraisal of progressiveness it may be necessary to study the development of theories in the medium-to-long term; for, research traditions may take time to show their mettle.

On this widespread view, of which Laudan is a keen advocate, the rational reconstruction of scientific progress requires, therefore, that rival theories from different research traditions are relatable, that the role and function of a theory within some research tradition can be explicitly recognised, and that the various stages in the evolution of a research tradition can be characterised and compared.

On Laudan's account, research traditions are the relatively stable entities whose progressiveness is to be assessed. By contrast, individual theories are transient and ephemeral beings whose cognitive appraisal depends, among other things, on background assumptions and on their interaction with competitors. In a rational reconstruction, however, the situation is likely to be reversed. Theories are the objects which can be reasonably precisely defined and delineated, wheres research traditions, as complex, open-ended and evolving entities, are much harder to pin down with any precision.

To any research tradition, a collection of theories can be associated. But the tradition amounts to more than this 'sum of its parts', because it generally contains background doctrine; especially ontological assumptions about the basic domains of scientific theories and methodological norms and commitments that direct the lines of inquiry and assess the results. Laudan summarises these additional components as comprising: "(i) a set of beliefs about what sorts of entities and processes make up the domain of inquiry and (ii) a set of epistemic and methodological norms about how the domain is to be investigated, how theories are to be tested, how data are to be collected, and the like" (Laudan, 1981a). Further, he holds that research traditions cannot be directly tested: the progressiveness of a tradition is a matter of the collective progress exhibited by its constituent theories as measured by their problem-solving effectiveness (PSE). But clearly the tradition establishes constraints on acceptable theories and is instrumental in weighing the relative importance of the problems they solve. In this sense it influences, of course, the judgement of rationality.

What do the various theories making up a research tradition have in common, apart from their being declared to belong to such-and-such a tradition? Evidently, they all conform in some way to those additional ontological and methodological presuppositions. But how is this conformity revealed at the level of *theory*, when a 'theory' is understood to be an exact structure of the kind characterised in the last chapter? One could, in principle, try to provide an exhaustive list of the background assumptions of a research tradition, and then seek a rigorous transcription of them into our metascientific framework; but this does not strike me as the most fruitful way to proceed here. What I propose instead is that we exploit the fact that the collection of theories comprising a given research tradition should display certain family resemblances; in short, it should be a *homogeneous* set. What this means exactly can be left open for the moment. The essential point is that the members of this collection of theories should be *interconnected*, and my suggestion is that as a first approximation we use the relations in ITR to describe this property. In other words, we represent the research tradition simply as a relational structure of the form

$$\mathbf{T}^* = \langle \Pi^*, \alpha_1, \dots, \alpha_k \rangle$$

where Π^* is the appropriate collection of theories, and $\alpha_1, \dots,$ α_k belong to ITR.

This description has the advantage of being simple and direct. Rather than construing the cognitive content of a research tradition in terms of its set of theories *plus a set of external assumptions and constraints*, we can think of the content of the tradition as being reflected in the *relational structure*, or in the way in which the various theories *fit together*. Whether this is an accurate and useful description, and whether it embodies all the distinctive features of the research tradition, is an open question that has to be looked into from case to case. Much depends on which particular relations α_i are included in $\mathbf{T}^*$ and what background assumptions are operative within the tradition. If necessary we might wish to expand the set ITR of admissible relations in order to obtain a more realistic representation. However, it is important to notice that this description is already quite informative: on the one hand, it captures the cognitive connections between the different theories of the tradition, and, on the other hand, it shows that some theories — actual or potential, conflicting or compatible — will be excluded from the tradition if they cannot be integrated into the given structure.

One problem with a system like $\mathbf{T}^*$ is that it may turn out to be too large or too amorphous to form a workable cognitive unit. This is likely to be the case if the research tradition in question is long-lived and comprehensive. We can make the structure more manageable by focusing on some particular stage in the evolution of the tradition. As a rule, not all theories of a research tradition are mutually consistent, and in a given domain there may exist rival, incompatible theories. But if the tradition is reasonably successful in resolving conceptual problems like inconsistency, and if we consider only collections of theories that scientists are willing jointly to assert during some period, then the resulting structure may be taken to comprise a domain of mutually compatible theories. Let us add the assumption that each of them can be ascribed a common, adequate logic in the sense of Chapter 4.

The next step towards obtaining a stronger grip on the structure of the research tradition consists in discarding from T^* those relations among $\alpha_1, \ldots, \alpha_k$ that are too weak, for instance the relations α and γ (if they occur) which typically hold between theories that are rivals. Let us furthermore agree to consider not the entire set of theories asserted within the tradition at the stage in question, but only a central 'core' of theories that are connected by *one* of the strong intertheoretic relations in ITR. Context may make it clear which core of theories should be chosen and which relation is the appropriate one. In general, the selection can be based on precise criteria. Among them, the following two are important: prefer relation α_i to relation α_j if $\alpha_i > \alpha_j$, and limit the domain of theories to those which are partially ordered by the preferred relation.[5] In this manner, we are led to restrict our attention to a system

$$T = \langle \Pi, \leqslant \rangle,$$

where T is now a substructure of our original T^* (after some relations have been removed). The domain Π of T then consists of an interrelated core of theories that are somehow 'central' to the research tradition at the chosen time, and these theories are mutually consistent and share a common adequate logic relative to which the ordering $\leqslant$ is definable.

More homogeneity can be imposed by requiring that T is *connected* in the sense that every pair of theories in Π has either a *meet* or *join* (under $\leqslant$) in Π. If this were not the case, T would possess distinct 'parts' perhaps only connected by weaker relations, and we would be more inclined to study those parts as separate structures. If each pair of theories has a join (respectively: meet) in Π, then **T** is a *join* (respectively: *meet*) *semi-lattice* containing a greatest (respectively: least) element. In that case, this core of theories belonging to the research tradition also possesses a 'core' element which we may term the *base theory* of **T**, and all other members of Π are related to the base by $\leqslant$. For a structure of this kind, Veikko Rantala and I have proposed the name: *theory ensemble*.[6]

My suggestion, in brief, is that theory ensembles can be taken as basic units in reconstructing research traditions, their evolution, and their interaction with rivals. To this end an ensemble, to be fruitful, should contain a sufficiently 'rich' domain of theories if it is to be a reasonably faithful characterisation of the research tradition at a given moment. Despite the apparent strength of our assumptions, there seem to be examples from the history of science where at least the kernel of a research tradition can be captural by an ensemble ordered by one of the stronger inter-theoretic relations, expansion, ε, or extension, σ (or their duals, ε', σ'). When this is not possible, one may try to base the ensemble on a weaker relation, like embedding or interpretation, or to represent the tradition at any stage by two or more ensembles. The latter situation would give rise in a natural way to 'higher-order' structures which have ensembles as elements.[7] However, to simplify matters, I shall assume here that an ensemble ordered by a strong relation represents a research tradition at a given time, or during a longer time interval if the theories in question are stable ones.

Suppose, then, that $\mathbf{T} = \langle \Pi, \leqslant \rangle$ is a theory ensemble whose ordering is defined by expansion, ε. Visually, we can depict **T** (as in Figure 3) as consisting of a base theory T, say, together with various expansions of the base; where T_i, T_j, ... contain addi-

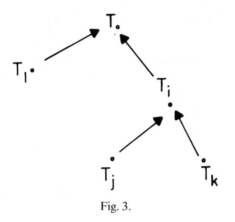

Fig. 3.

tional concepts and laws, besides those of T. But by the property that T_i, T_j ε T, there is a translation of these concepts and laws back into the base. If the ordering is defined by the extension relation, σ, the picture is similar, except that now the base can be imagined as a conceptually 'comprehensive' theory, and T_i, T_j, . . . express additional laws using only a part of T's vocabulary.

When the 'dual' relations, ε' or σ', characterise the ordering of the ensemble, it is more natural to think of **T** as a meet semi-lattice. The picture is then inverted, with the base theory T appearing at the bottom of the lattice, and with the property that $T \leqslant T'$ for any theory T' in Π (see Figure 4). For example, if $\leqslant$

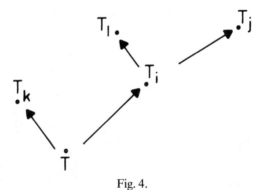

Fig. 4.

is the relation σ', then $T \sigma' T_i, T_j$; which means that T_i and T_j have additional concepts, but with no requirement this time that they are translatable into the framework of the base theory. Rather, by the property $\sigma' > \delta$, there is an interpretation, and hence translation, of the *base* into $T_i, T_j, \ldots$

An exemplary type of theory ensemble can be extracted from the Newtonian tradition in mechanics. The core of this tradition can be taken to comprise classical (point) particle mechanics (CM) and its application to celestial, Hooke's law, electrostatic systems, and so forth. CM, the base theory of this core, will have Newton's second law of motion as its central axiom. The remaining theories can be thought of as expansions of the base in which additional concepts of force are introduced along with new terms denoting physical constants appropriate to each of the types of system or problem domains studied, for example field potentials, dialectical constants, and so forth. Each of these further theories presupposes or entails Newton's second law, yet it seems that in every case the new concepts they deal with are in some sense definable from the primitives of CM so that there is a translation back into the base theory and the collection as a whole forms an ensemble ordered by expansion.

We should not anticipate that all mature theories in the Newtonian research tradition can be incorporated in a single ε-ensemble. To include, say, continuum mechanics or geometrical optics we should have to construct further ensembles, or to weaken the central intertheoretic relation in the existing structure; for instance, a larger domain of theories could be handled by taking embedding rather than expansion as the basic relation. However, Balzer and Moulines (1981) have shown how as many as fourteen different theories of classical mechanics can be accommodated within one semi-lattice structure, and it would seem to be a reasonable conjecture that this structure or something very like it can be reconstructed as an ε-ensemble. One of the main advantages of the ensemble representation over Balzer and Moulines' 'theory nets' is that, unlike in their case, we do not require the base to be a 'conceptually rich' theory containing all of the additional terms needed for the various applications of the central force law; this would in fact make the base theory a highly artificial construct, rather than a genuine physical theory actually held by scientists working within the Newtonian tradition.

The applicability of the ensemble representation is not confined to mechanics or even to physical theories. To give one example, Michele Tucci and I (1982) have suggested how complexes of interrelated theories from mathematical economics can also be pictured in this format.[8]

The fact that an ensemble, especially if ordered by ε or by σ, is a tightly connected array of theories enables it to convey a good deal of the relevant background information pertaining to the research tradition. Suppose that the core of a tradition is captured by an ensemble, $\mathbf{T}$. Then consider, for instance, the possible ontological and linguistic assumptions of the tradition which specify the basic domains of scientific inquiry and the fundamental concepts to be employed in investigating those domains. Let the base theory of $\mathbf{T}$ be $T = \langle \tau, N, M, R \rangle$. The type τ of T then lays down the basic ontological and conceptual categories for the whole ensemble, and thus for this core of the research tradition. Each model of T comprises a universe of discourse corresponding to each sort in τ, and a concept (predicate, function, etc.) corresponding to each symbol in τ. Moreover, any other theory T' in $\mathbf{T}$, has a type τ', say, which includes all sorts and symbols from τ; and, by the property that T' is an expansion of T, there is a translation of τ'-sentences into τ-sentences. Now, for generality's sake I have described translation as a sentence-to-sentence mapping. But, in typical cases where the translation is built up recursively from atomic formulas, the domains and relations corresponding to the extra sorts and symbols in τ'

should be in some way constructed or defined from those of τ. And naturally this means that there is a certain ontological and linguistic uniformity present throughout the entire ensemble, even if individual theories (especially those located at the extremes of different chains in the lattice) may differ markedly from one another.

Aside from ontological and linguistic commitments, also methodological and normative aspects of a research tradition can be reflected in the ensemble schema. Examples of normative constraints might be the following: try to develop a new theory under the restriction that it be absorbed into an existing ensemble; try to merge distinct parts of a tradition into a single ensemble; search for new or improved theories whilst aiming to keep intact as much as possible of an established ensemble, preserving say the base or some central part of the structure.[9]

In this respect the role of 'R' may be particularly important. Take the previous ensemble **T** with base T, and suppose that R represents symmetries or invariance properties of T. Though I have not included an explicit treatment of the component 'R' in intertheory relations, there are many cases where it is natural to require that R be in a suitable sense preserved under relations like ε and σ.[10] If so, all theories of **T** would possess some common symmetry properties. A paradigmatic example of this arises, once again, in the Newtonian tradition in mechanics, where a constraint on acceptable theories would be that their laws should be invariant under Galilean transformations. The fact that Maxwell's field equations did not share this property represented, of course, an anomaly for the Newtonian tradition, or perhaps, to use Laudan's terminology, a *conceptual* problem. This we might formally represent as the impossibility of integrating Maxwell's electrodynamics into an ensemble based on classical mechanics. As it turned out, this anomaly spawned the rise of a new, relativistic, research tradition in mechanics in which the Lorentz invariance of Maxwell's theory was to become a key normative constraint on the development of new theories.

4. THEORY CHANGE AND RELATIONS
BETWEEN ENSEMBLES

I began by introducing the ensemble as a chiefly static structure, but considerations of the ways in which it might reproduce certain broader features of the research tradition led quickly and naturally to a discussion of theory change. In general, to achieve a diachronic representation of theories we shall have to match each stage in the life of a research tradition to a distinct ensemble or set of ensembles (again I shall assume, for simplicity, that one will suffice). In this manner the tradition as a whole can be depicted by a sequence of structured collections of theories.

On Laudan's account, research traditions progress by improving the problem-solving effectiveness of their constituent theories. This can happen if established theories are modified to solve new empirical problems; if new theories are found for the same purpose, or for solving anomalies occurring when a previously unsolved problem has been satisfactorily dealt with by a rival tradition; or when residual conceptual problems are suitably resolved. If $T = \langle \Pi, \leqslant \rangle$ is an ensemble characterising the core of the research tradition at a particular time, then progress in any of these senses will involve a combination of expanding the domain Π by inserting new theories, and replacing or modifying some of the members of Π. This will give rise to a new ensemble, say $T' = \langle \Pi', \leqslant' \rangle$. In addition, if the transition from T to T' marks a progressive evolution of the research tradition, we should expect the following conditions to be fulfilled:

(i) $PSE(\Pi') > PSE(\Pi);$

(ii) $\Pi \cap \Pi' = \Pi_0 \neq 0;$

(iii) $\leqslant \upharpoonright \Pi_0 = \leqslant' \upharpoonright \Pi_0.$

The first simply requires the combined problem-solving effectiveness of Π' to exceed that of Π. The others are continuity conditions requiring that T and T' have overlapping domains of theories and that they order this shared subdomain in like fashion, i.e. $\langle \Pi_0, \leqslant \upharpoonright \Pi_0 \rangle$ is a common substructure of T and T'.

Unlike Kuhn and Lakatos, Laudan does not insist that some 'inner core' of assumptions or theories survive the entire life span of a research tradition. Consequently, there is no simple criterion which would determine when a tradition has effectively expired, and even a stagnating or regressive tradition might eventually show signs of revival. In general, therefore, (ii) and (iii) seem to be the strongest requirements of this type that we can impose on the relation between temporally adjacent ensembles. However, a characteristic of 'decline' might be that persistent attempts to enlarge the domain of an ensemble fail to produce positive results in terms of increasing PSE; or that development of the tradition is bought at the cost of fragmenting an extant ensemble by splitting it into disjoint structures; since this would tend to indicate the presence of conceptual problems. Naturally, the more changes in Π that occur in the transition from T to T', the more radical the development of the tradition. At the other extreme, if $\Pi \subset \Pi'$, the tradition may be said to exhibit *cumulative* progress.

In general, progress is not merely a matter of the internal changes taking place within a research tradition: as we have seen, theory appraisal is taken to be largely a *comparative* affair whereby the relative performances of two or more competing theories, possibly from rival traditions, are evaluated. So, to achieve a satisfactory index of progressiveness, we must be able to keep track of the relation not only between an ensemble T and its successor, but also between T and an ensemble representing some contemporary competitor, call it T'. Let us suppose once again that relations in ITR are suitable for this purpose. But, this time, since T and T' are deemed to be rivals, it may be safer to assume that the weaker relations like (limiting case) correspondence (or perhaps interpretation and embedding) are more appropriate.

What does it mean, then, to say that there is a *correspondence* of an ensemble $\mathbf{T} = \langle \Pi, \leqslant \rangle$ to an ensemble $\mathbf{T}' = \langle \Pi', \leqslant' \rangle$? Obviously, theories in Π must be in correspondence with counterparts in Π'. I think it would be a mistake to require generally that *all* members of Π are thus related to some theory of $\mathbf{T}'$, just as there may very well be elements of Π' which possess no corresponding theory from $\mathbf{T}$. In short, a correspondence of $\mathbf{T}$ to $\mathbf{T}'$ should be a partial function f on Π' that assigns to each theory T in its domain a pair $\langle f(T), \langle F, I \rangle_{T, f(T)} \rangle$, where $f(T) \in \Pi$ and $\langle F, I \rangle_{T, f(T)}$ is a correspondence of $f(T)$ to T relative to $\langle L, L' \rangle$ (being respectively the logics adequate for theories of the two ensembles). A second condition for correspondence to fulfil is that the mapping f should be a *morphism* of the posets in the sense that for any theories T, T' in the domain of f,

$$T \leqslant' T' \Rightarrow f(T) \leqslant f(T').^{11}.$$

Other inter-ensemble relations can be defined analogously. Naturally, if f is say an *interpretation of* $\mathbf{T}$ and $\mathbf{T}'$ it will assign to a given pair T, $f(T)$ an interpretation of $f(T)$ in T in the obvious way.

In some contexts it is desirable to add further requirements on the map f; a natural one being to ensure that in some sense the 'same' correspondence, interpretation or whatever is linking theories in Π with their counterparts in Π'. This idea can be illustrated with respect to the interpretation relation. First, notice that if $\mathbf{T} = \langle \Pi, \leqslant \rangle$ is an ensemble ordered by ε, and T_i, T_j are any two theories in Π with $T_i \leqslant T_j$, there is a unique translation, I_{ij} say, from sentences in the language of T_i into sentences in the language of T_j. In particular, if T_m, $T_n \in \Pi$, and $T_i \leqslant T_m \leqslant T_j$ and $T_i \leqslant T_n \leqslant T_j$, then $I_{ij} = I_{mj} \cdot I_{im} = I_{nj} \cdot I_{in}$; in other words, the following translation diagram commutes,

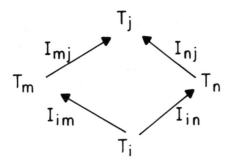

Now let $\mathbf{T}' = \langle \Pi', \leqslant' \rangle$ be an ε-ensemble such that $\mathbf{T}$ is interpretable in $\mathbf{T}'$ by a partial function $f: \Pi' \rightarrow \Pi$. We want to express the property that the interpretation is *uniform* across the different theories in $\mathbf{T}$. What we know already is that for any T, T' in the domain of f, $T \leqslant' T' \Rightarrow f(T) \leqslant f(T')$. We can therefore use properties of the translations assigned by f to ensure that the interpretation of $f(T)$ in T matches that of $f(T')$ in T', namely that the following condition holds:

$$I_{TT'} \cdot I_{f(T)T} = I_{f(T')T'} \cdot I_{f(T)f(T')}.$$

This implies that the result of translating a sentence φ in the language of $f(T)$ into a sentence in the language of T' is the same whether we first interpret $f(T)$ in T and then apply the translation of T into T' or whether we first express φ in the language of $f(T')$ and then apply the interpretation of this theory into T'. In this manner we are able to ensure that the different translations assigned by f mesh together in the appropriate way.

5. THEORY CHANGE AND CONTINUITY

We are now ready to consider in more detail how the problem-solving ability of a research tradition might be matched by that of a rival. To begin with, suppose that the core of the tradition in question is characterised by an ensemble $\mathbf{T} = \langle \Pi, \leqslant \rangle$ whose ordering is defined by the expansion relation. Let $T = \langle \tau, N, M, R \rangle$ be the base of the ensemble, and assume that the class of models M of T is defined by a single axiom $\theta \in \text{Sent}_L(\tau)$.

Problem-solving in this tradition can therefore be represented by the process of enriching the vocabulary of T, formulating suitable additional laws in the expanded language, and using these together with appropriate initial conditions to derive statements expressing the relevant problems. Let $T_1 = \langle \tau_1, N_1, M_1, R_1 \rangle$ be an expansion of T in $\mathbf{T}$ which solves in this manner a problem p expressed by $\varphi \in \text{Sent}_L(\tau_1)$. Let θ_1 be the axiom of T_1, and suppose that ψ formalises the initial conditions; so that

(3) $\theta_1, \psi \vDash_L \varphi.$

Now, the fact that T_1 is an expansion of T implies that p can also be expressed in the framework of T, and indeed that the solution given by (3) can in a certain sense be reduced to a corresponding solution in the language of T. For, let $\langle F_1, I_1 \rangle$ define the expansion relation holding between T_1 and T. We have:

$$F_1 : M_1 \restriction \tau \rightarrow M_1;$$
$$I_1 : \text{Sent}_L(\tau_1) \rightarrow \text{Sent}_L(\tau);$$

where, for each model $\mathfrak{M}$ in the domain of F_1, $F_1(\mathfrak{M}) \restriction \tau = \mathfrak{M}$, and the restriction of I_1 to L-sentences in the type τ is an identity mapping. Plainly, the additional relations in τ_1 are implicitly defined in M_1. If we add the plausible assumption that they are also *explicitly* defined, and that the translation I_1 is determined according to explicit definitions of the $(\tau_1-\tau)$-terms, it follows that the sentence $I_1(\theta_1)$ axiomatises the domain $M_1 \restriction \tau$ of F_1, i.e.

$$\text{Mod}_L(I_1(\theta_1)) = M_1 \restriction \tau.$$

Then, by the properties of expansion, it follows that the schema

(4) $I_1(\theta_1), I_1(\psi) \vDash_L I_1(\varphi)$

will represent a solution of p reproduced within the framework of T.[12]

Imagine now that a rival research tradition is depicted by an ensemble $\mathbf{T}' = \langle \Pi', \leqslant' \rangle$ also ordered by expansion. Let there be a correspondence f of $\mathbf{T}$ to $\mathbf{T}'$. Suppose too that the base $T' = \langle \tau', N', M', R' \rangle$ of $\mathbf{T}'$ is in correspondence with T. Though the definition of f does not formally require this to be the case, the assumption itself is reasonable. Thus f associates T· with T' and assigns some correspondence $\langle F', I' \rangle$ of T to T'. By hypothesis,

$$F' \colon K' \to K,$$
$$I' \colon \mathrm{Sent}_L(\tau) \to \mathrm{Sent}_{L'}(\tau'),$$

where K and K' are nonempty subclasses of M and M' respectively.

If we suppose that the class $M_1 \upharpoonright \tau$ is included in the range K of F', then we can define an embedding $\langle F, I \rangle$ of T_1 in T':

$$F \colon K'' \to M_1,$$
$$I \colon \mathrm{Sent}_L(\tau_1) \to \mathrm{Sent}_{L'}(\tau').$$

We set $K'' = \{ \mathfrak{M} \in K' \colon F'(\mathfrak{M}) \in M_1 \upharpoonright \tau \}$ and $F'' = F' \upharpoonright K''$, and define

$$F = F_1 \cdot F'',$$
$$I = I' \cdot I_1,$$

It is then a routine matter to check that F maps K'' onto M_1 and that I preserves truth relative to F. The upshot is that the problem solution (3) can be seen to be transposed into the framework of $\mathbf{T}'$. For, let K' be axiomatised by an $L'(\tau')$-sentence, σ'; i.e. $K' = \mathrm{Mod}_{L'}(\sigma')$. We thus obtain

(5) $\sigma', I(\theta_1), I(\psi) \vDash_{L'} I(\varphi)$.[13]

What consequences does this have for the problem-solving capacity of T' or its research tradition, represented by $\mathbf{T}'$? In the first place, notice that since the conjunction of θ_1 and ψ is consistent with θ, the conjunction of their translations $I(\theta_1)$ and $I(\psi)$ is consistent with σ'. Thus, from a purely logical point of view there can be no objection against using these statements as additional assumptions, together with σ', to obtain an answer to the problem under scrutiny. But it is important to notice too that (5) has all the makings of a 'borrowed' solution, handed down from the rival tradition. It seems that in order for (5) to count as a genuine solution to p, the conjunction

$$\sigma'' = I(\theta_1) \wedge \sigma'$$

should be an accepted law of $\mathbf{T}'$, or derivable from such a law. If this is not already the case, the strategy of the parent research tradition seems clear enough: if p is regarded as a significant problem, one should try to formulate an appropriate expansion of T' whose axiom is either strong enough to ensure the truth of σ'', or else independently implies $I(\varphi)$. This means finding a conceptual enrichment of τ' appropriate to the given problem domain, and proposing and testing new laws that are relevant there. The validity of schema (5) already indicates, however, that a measure of continuity exists between the two ensembles or their research traditions; just the kind of continuity, in fact, that is missing from Laudan's account and which his theory of progress clearly requires.

(5) is available, as we saw, under two key assumptions (apart from the initial hypothesis that the research traditions can be represented by an ε-ensemble).[14] The first is that $\mathbf{T}$ is related to $\mathbf{T}'$ by a correspondence which in particular connects the two base theories, T and T'. The second is that the range of the structural part of this correspondence includes the class $M_1 \upharpoonright \tau$; in other words, the class which is basic to the solution of p within T.

The first assumption is hardly problematic. In the present framework a correspondence f between the ensembles is a minimal requirement for establishing a reasonable problem-solving connection between the research traditions; and the most natural conjecture in this respect is that f relates T with T'. But little alters if in fact T is in correspondence with some other member of $\mathbf{T}'$; since that theory will also be reducible to T', and hence any transfer of problem-solving ability can ultimately be characterised in terms of the base of $\mathbf{T}'$.

As to the second assumption, we can generalise it somewhat by considering the ensemble $\mathbf{T}$ in its entirety. Each theory T_i in Π expands the base T and is, in part, characterised by a class of models, M_i say. In every case $M_i \upharpoonright \tau$ is a subcollection of M representing so to speak the problem domain of T_i seen from within T. Probably not all such theories will be deemed successful from the standpoint of the rival tradition; the latter will try to improve upon as well as reproduce the performance of $\mathbf{T}$. But we might think of the range K of the structural correlation F as capturing that part of T which is reckoned 'successful' by the lights of $\mathbf{T}'$. This is the part which in a sense T' 'retains' in the shape of the domain K'' of the correspondence. It follows that, to the extent that p is thought to be adequately solved by $\mathbf{T}$, we can anticipate that the class $M_i \upharpoonright \tau$ really is included in K; and likewise for the general case $M_i \upharpoonright \tau$.[15]

Considerations of this kind seem to be born out in practice. Rantala and I (1984a; 1988) have shown how a (limiting case) correspondence can be defined between classical particle mechanics (CM) and special relativistic particle mechanics (RM); actually we reconstruct CM and RM as fragments of these theories that are suitable as bases for ensembles characterising classical and relativistic mechanics. Some of the details of this example will be recounted in Chapter 7. It suffices for now to mention that in this correspondence, the domain K' of the structural correlation consists roughly of those models of RM in which particle velocities and accelerations are 'very small' compared to the speed of light. K' thus represents the domain in which, from a relativistic perspective, CM proved most successful and where it could later be retained as 'approximately' valid. The correspondence itself is specified by a more precise formulation of the usual idea that the central laws of CM and RM 'converge' in the limit of low particle velocities v, i.e. as the ratio $v/c \rightarrow 0$, where c is the speed of light. It is well known that in several problem domains this condition is sufficient for obtaining classical equations as limits of corresponding relativistic equations. A relativistic ensemble can therefore be regarded as containing several expansions of RM that are in correspondence with suitable members of the classical ensemble based on CM; the relation being similarly defined in each case, with the f images of the relativistic theories possessing model classes whose 'successful parts', reduced to the language of CM, are included in K. It is also widely accepted that, in some cases, additional assumptions and limits are needed to obtain classical theories as approximations of relativistic counterparts. The successful problem domains of these theories should similarly fall within K, though extra conditions may be required to define the correspondence relation.

In this manner it is apparent how the relativistic tradition can be said to preserve an essential part of the problem-solving effectiveness of classical mechanics whilst making corrections and improvements that increase PSE at the same time. There are, of course, classical theories possessing no relativistic counterparts. Either we shall have to regard these as belonging to a part of the Newtonian tradition that is, from a relativistic standpoint, unsuccessful; or else we can grant their problem solutions to be approximately adequate but concede that not all of the problem-solving capacity of classical mechanics can be (at least on present evidence) absorbed into the framework of the later theory. The last conclusion is, of course, compatible with Laudan's analysis of 'progress'. It also supports our earlier decision not to insist that all theories in the 'reduced' ensemble be matched by corresponding members of the 'reducing' ensemble.

If Laudan and Kuhn are correct, the situation just envisaged is probably not untypical when the ensembles in question represent two radically different research traditions one of which comes to supersede the other through what amounts to a scientific 'revolution'. Scientists working on **T′** will reject out of hand some theories belonging to the rival ensemble **T**. Problems which these theories 'solved' may simply vanish or become 'meaningless' from the standpoint of **T′**. Classical theories of the aether, for example, had no place in the relativistic tradition. The empirical problems they were designed to answer were rendered obsolete and the anomalies and conceptual problems they generated received a satisfactory resolution. In other cases, however, the failure of **T′** to develop a suitable counterpart to some theory of **T** may be taken as a sign of inadequacy. Then, whether **T′** can still be regarded as the more progressive of the two may depend on other factors; for example, to what extent **T′** improves on **T** in its handling of their shared problems, and how the weights assigned to any new problems solved by **T′** compare to the weights given to any anomalies created by the inability to import all of **T**'s problem successes into the new framework.

Sometimes this failure to import some empirical successes may be a temporary, practical difficulty, later resolved by suitable adjustments to **T′**. But the problem may also be 'structural', and highlight an intrinsic property of the ensembles or the relations between them. Here is a hypothetical example. Suppose that **T** is interpretable in **T′**, and that I is the translation from the language of the base T of **T** into the language of the base $T′$ of **T′**. Let T_1 be an expansion of T in Π whose axiom translated into the language of T is $\sigma \in \mathrm{Sent}_L(\tau)$. It may happen that $I(\sigma)$ is incompatible with the axiom $\theta′$ of $T′$. This would imply that $I(\sigma)$ is incompatible with any expansion of $T′$. Thus, either T_1 cannot be interpreted in any member of **T**$'$ or else any such interpretation would not square with the translation I in the manner which I recommended earlier. In this situation we would have to conclude that T_1, and perhaps also its problem domain, is effectively excluded from the realm of **T**$'$; unless, of course, we were prepared to entertain a modification to the base theory T' or to redefine the logical relation of **T** to **T**$'$. This may well be unnecessary, however, if, as is likely, the theory T_1 is thought to be unsatisfactory from the standpoint of **T**$'$, and thus to present no anomalies for it.

6. ENSEMBLES AND THE PROBLEM-SOLVING MODEL OF PROGRESS

I want to conclude briefly by drawing together some of the loose strands and considering some of the broader implications of our analysis for the problem-solving account of progress and Laudan's model in particular.

We have seen, both earlier and in this chapter, how Laudan's approach to the issue of rational change in science is heavily indebted to the idea of problem-sharing between research traditions and grounded on the possibility of establishing contact and communication across different traditions. If the arguments of Chapter 3 are substantially correct, an impartial determination of problem-solving effectiveness, and hence a comparative evaluation of progressiveness, is barely feasible in the absence of any shared problems and common methods of assessing problem solutions. It is essential, therefore, for Laudan's account, and indeed for any viable theory of progress, to be able to keep track of the shared problems when they arise, and to keep open as far as possible the lines of communication between rival scientific research traditions.

In this respect, the notion of a theory ensemble embodies some of the key ingredients for characterising a research tradition as a homogeneous family of theories and for endowing it with some indispensible 'internal' structure. Correspondence and stronger logical relations between ensembles then serve to identify the principal lines of communication between traditions, and to indicate how problems and problem solutions may be transmitted from one tradition to another.

Our account of the evolution of traditions, and the logical relations that may hold between rivals, left PSE largely as an undefined primitive. Can we now go one better and unpack this progressiveness-count into more manageable components? This problem is a complex one, and I have no simple solution to hand. But the foregoing analysis can, I think, help us to make some headway and to improve on Laudan's own version of the events.

In the first place it seems that we can achieve a much clearer grasp of *anomalies* than that provided by Laudan's informal discussion. Recall that, for Laudan, a weighted aggregate of problems that are anomalous for a theory or research tradition represents a key negative component of PSE. For a problem to be anomalous it is not essential that its formulation be inconsistent with existing theories of the tradition; it suffices, in fact, that the problem has been adequately solved by a competitor. And, by the same token, a recalcitrant problem or even an 'empirical inconsistency' need not be regarded as anomalous for a theory if no serious competitor has adequately resolved it.

Consider, then, the case of two rival research traditions characterised as before by ensembles **T** and **T′**; and suppose for simplicity that they are the only competitors in a given field. On Laudan's view it should follow, therefore, that, leaving aside any shared problem successes, all of **T**'s solved problems will be anomalous for **T′**, and conversely (cf. Figure 2 once again). But this is surely too simplistic. Laudan himself seems to realise this, for he adds the rider that the theories or traditions in question should share the same domain. The trouble is, of course, that Laudan offers no suggestions for how the sameness of domain to be recognised. Our account seems to provide a way out of this difficulty, and leads to the following considerations.

First, to be anomalous for **T′**, a problem p solved by **T** under the schema (3), say, should be *translatable* into the framework of **T′**. A problem that cannot even be expressed in the context of **T′** can hardly pose a cognitive threat to its research tradition; it simply lies outside its scope. If, as some historians and philosophers believe, the phlogiston theory is wholly or partly untranslatable into the language of Daltonian chemistry, the fact that not all problems 'solved' by that theory might be transformed into empirical successes for modern chemistry would scarcely be a cause for alarm. The presence of a correspondence between **T** and **T′** is thus a first step in clarifying whether and how a given problem may be available under translation in both research traditions.

Still, translation may be necessary but not yet sufficient for construing an unsolved problem as anomalous. Any problem which, from the vantage point of **T′**, lies outside the domain where **T** is at least approximately adequate will be similarly innocuous. Even if the central terms of the phlogiston theory do admit of translation into a later idiom, not all of its problems will remain significant; some just vanish into obscurity, most likely branded as nonsensical. For this reason the inclusion of the structural correlation F as part of the correspondence relation is important. If the domain of F is interpreted, as suggested, as representing within **T′** a successful part of **T** (or its base, T), we may think of this as delineating a class of problems that are not only translatable into the framework of **T′** but are potentially anomalous for that research tradition.

If, under such a translation the formulation of the problem is consistent with **T′** (as in the case of schema (5), above), then the problem has a potential solution within that tradition, and indeed some guidelines for approaching a solution may be forthcoming. If unsolved, it remains temporarily anomalous, nonetheless, and can play a role as such in the evaluation of PSE for the tradition. If the transformed problem is incompatible with one or more theories of **T′**, the situation is more complicated. The incompatibility may arise already at the level of the base theory *T′* and thus conflict with some fundamental assumptions of the tradition which are less likely to undergo piecemeal revision. Or incompatibility may occur at some other place in the ensemble where perhaps scientists are more willing to concede possible modifications to accommodate a successful resolution of the problem. In the former case, the problem might well be ignored as being inadequately solved, or improperly posed, by **T** in the first place. In the latter case, the anomaly might be transformed into a solved problem, bringing additional credit to **T′** or its successors. Alternatively, one might be led to redefine the connections between **T** and **T′** and thereby reinterpret the problem itself.

These considerations show, I believe, how our formal meta-scientific framework may bring us closer to an accurate evaluation of problem-solving effectiveness. They lead to a generalisation of Laudan's analysis of empirical problems and anomalies, schematised in Figure 5. There remains, of course, a number of open questions: How can empirical problems be counted and weighed for their importance, and how can conceptual problems be individuated and assessed? Again I think the ensemble representation may have some value here. I do not wish to pretend that there is an obvious and straightforward method of 'counting' problems and arriving at a quantitative measure of PSE. What we have for each research tradition is an interrelated collection of theories in the form of an ensemble. We may regard each theory as dealing with a single problem-type or 'cluster of problems', or a related set of distinct problems, according to more general

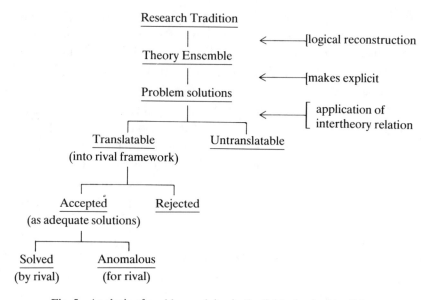

Fig. 5. Analysis of problem-solving in the light of a rival tradition.

pragmatic criteria, or according to taste. Still, each member of the ensemble can be regarded as a problem-solving unit, and analysed as such. Which of the solved problems are more important, and thus carry greater weight, is also a pragmatic question. But the structure of the ensemble may provide some clues in this direction. The base theory is, of course, fundamental to the ensemble and in a sense 'supports' the entire structure. Likewise, those theories situated immediately below (or above) the base support fragments of the remaining structure, and so on for theories below (or above) them. Theories which in this manner support long chains of elements, or theories from which many different chains 'branch out', evidently play a more central role in the research tradition, and the problems that they (or the elements they support) solve may carry extra weight accordingly.[16]

The conceptual problems of a research tradition are also

reflected to a large extent by formal properties of the ensemble scheme. Most of the conceptual problems which Laudan discusses can be reduced to problems of incompatibility between, or a lack of sufficient homogeneity among, the various theories of the tradition. Both types of problem have formal analogues: compatibility is achieved when as many theories as possible can be collected into a single structure; homogeneity is assured when that structure can be ordered by one of the stronger intertheoretic relations. This is, of course, only a crude and preliminary analysis of conceptual problems, and of PSE in general; but perhaps these remarks will give some hint of the direction in which future research on this topic should proceed.

6

MEANING CHANGE AND TRANSLATABILITY

1. MEANING AND CONCEPTUAL CHANGE

The hardcore philosophical issue raised by the doctrine of incommensurability, and the locus of most of the philosophical wrangling it has provoked, is the problem of the *comparability* of rival theories and the *rationality* of theory appraisal and choice. That competing theories are *to some extent* comparable is not, of course, at issue. The conflict between Kuhn, Feyerabend and their critics concerns the precise nature of that comparison. Can the *cognitive contents* of rival theories be rationally compared? If so, by what criteria of rationality? It is well known that Kuhn and Feyerabend offer up different and rather complex answers here. Very crudely put, Kuhn's response is in favour of rational comparability, but at the expense of adopting criteria of rationality that depart from the prevailing norms of empiricism and scientific realism to a degree that many rationalist philosophers find intolerable. Feyerabend's response amounts to a critique of the very concept of scientific methodology and rationality, and a plea for the meta-methodological principle 'anything goes'.

In recent years, most of the philosophical discussions of theory change and rationality in science have been centred on the problem of the alleged meaning variance of scientific concepts. Many writers have argued, along with Kuhn and Feyerabend, that the central terms of a scientific discipline acquire their meaning wholly or substantially in the context of a given theory or theories. As these evolve in the course of time, so meanings alter accordingly. And when radical changes occur in the conceptual scheme of a science, such as in revolutionary changes of paradigm, the shift in meaning that a given term undergoes may prevent a straightforward comparison of the concept that it names in the old paradigm with the new meaning that it subsequently acquires. When this kind of meaning variance is sufficiently deep-seated and extends to all the primitive terms of a given theory, it is often said to result in semantic incommensurability, and nontranslatability. The whole question of rational appraisal and choice becomes problematic in such cases, since the cognitive content of the earlier theory is not re-expressible in the language of the new; and, in particular, it is held that crucial experiments cannot be invoked to provide unequivocal empirical grounds in favour of the new theory, because there is no common, neutral language in which such experiments could be described.

This situation is often termed "the paradox of meaning variance". The 'paradox' in question arises from the fact that *prima facie* incompatible, rival theories may display such radical meaning changes that it becomes no longer possible to claim that they are using, and referring to, the 'same' concepts and quantities. Since their respective laws (consequences, predictions, and so forth) cannot then be brought into genuine logical relations — of entailment, inconsistency, etc. — with one another, it appears that empirical observations cannot be used to choose between them, even though scientists normally act as if experiments were decisive.[1]

There are many examples of alleged meaning variance of this kind; none more widely discussed and more hotly disputed than that of the concept of *mass* in classical mechanics and its relation to the term of the same name in special relativity (and in post-relativistic physics generally). Though I wish to consider the problem of meaning variance and incommensurability in a fairly wide context, it will be useful in both this chapter and the next to keep this example in the foreground and to treat it as something of a test case.

Consider first one of the standard formulations of the relation of classical mechanics (henceforth CM) and special relativistic mechanics (RM). CM and RM are taken to share a common stock of primitives: mathematical and spatio-temporal terms, together with the terms 'mass' and 'force', belong to each theory. RM has, additionally, an individual constant c denoting the velocity of light *in vacuo*. And if rest (or proper) mass is construed as a primitive of RM, then relativistic mass will be a defined term, in the usual way. Hence, the vocabulary and syntax of CM is contained in that of RM, and the two theories are mutually incompatible — the Minkowski force law of RM being incompatible with Newton's second law of motion. Despite the disagreement in their central laws, the predictions of the two theories accord well over a limited range of phenomena. This property can be given a mathematically precise formulation by saying that several relativistic equations pass asymptotically into classical laws when suitable limits are approached, most importantly the limit $c \rightarrow \infty$, or the ratio $v/c \rightarrow 0$, for particle velocities v.

The mathematical details of this *correspondence* between CM and RM are rather well known and are in any case rarely challenged by disputants in the commensurability debate. They are relevant, however, only to the syntactical part of the story. The crucial semantical part of the tale remains to be explored. What, then, are the conceptual relations between the theories?

There is a popular fallacy that according to logical empiricism the meanings of terms like 'mass' remain stable in the transition to a new theory. Many commentators have thus regarded the doctrine of meaning variance (even of the less radical sort) as a criticism of empiricist and especially positivist accounts of science. The popular view should be qualified, however. Long before the rise to prominence of the incommensurability thesis, logical empiricists such as Philipp Frank had in fact argued that, though mutually inconsistent (and thus comparable), CM and RM

were to some extent conceptually disparate. Classical and relativistic mass, for instance, have different meanings because certain sentences required to be true in order for laws involving
Newtonian mass to be empirically interpreted are actually false if
relativity theory is deemed to be correct. This implies that from
the standpoint of RM Newtonian 'mass' is at best ambiguous, and
at worst completely uninterpreted, i.e. without empirical denotation.[2] There are more modern variants of this argument, less
'operationalist' in spirit, which appeal to considerations of *measurement*. Quantities like mass must be treated as *theoretical*, to
the extent that in order to measure them in a given context one
has to presuppose the validity of certain statements that amount
to theoretical laws. For example, in the case of Newtonian mass,
such measurements seem to require the validity of Newton's
second law, or the law of conservation of momentum. When these
statements contradict their counterparts from another theory (as
may be the case with RM) there is bound to be a conceptual
discrepancy between the terms denoting this quantity in the two
theories.[3]

The thesis that meanings change with theoretical context does
not therefore conflict with traditional, empiricist views of science.
But the stronger claim associated with the incommensurability
thesis does conflict — namely, that conceptual disparity affects *all*
terms of a theory in such a manner that no common basis for
logical comparison remains. For Kuhn, Feyerabend and others,
meaning variance persists right along the conceptual spectrum,
from the more theoretical to the more observational terms. It is
not that 'mass', for example, happens to be differently defined in
CM and RM from shared underlying concepts, but rather differently defined *tout court*. Moreover, such conceptual differences
are held to be so radical as to rule out any possibility of
translation between the two languages. Even Kuhn, who has
vacillated on occasions between weaker and stronger formula-

tions of this claim, has recently reaffirmed in no uncertain terms the nontranslatability thesis in connection with CM and RM (Kuhn, 1983). His holistic account of meaning is equally uncompromising. Having conceded the possibility in principle that *some* meaning-preserving terms could provide a basis for the rational comparison of theories, he proceeds to scotch the very same idea a few sentences later:

> It is simply implausible that some terms should change meaning when transferred to a new theory without infecting the terms transferred with them.   (Kuhn, 1983)

One of the stock replies to the thesis of radical meaning variance and incommensurability has been disbelief and flat denial. According to some authors, no problems of conceptual disparity or nontranslatability arise in the case of CM and RM because terms like 'mass' denote the *same* concept in each theory. In both classical and relativistic mechanics, inertial mass is the property of a body in virtue of which it resists a change in motion. The validity of the conservation of momentum law in both CM and RM, and the fact that force is always defined as rate of change of momentum, are taken to indicate that the two notions of mass, classical and relativistic, really amount to one and the same concept. However, it is hard to construct a plausible defence of this position. In particular, if Newtonian mass is defined as relativistic rest mass, different expressions are obtained for momentum and force in the two theories, and hence different criteria for measuring mass apply. On the other hand, the identification of classical and relativistic mass *simpliciter* has to live with the fact that the former but not the latter is velocity or frame independent.[4]

2. STABILITY OF REFERENCE

There is another type of response to the incommensurability

thesis which has found an especially favourable reception among philosophers with a leaning towards scientific realism. It resides in the claim that as a rule rival theories are *referentially* comparable. In plain words, even if some central terms of a scientific theory change *meaning* (or Fregean *sense*), there will usually exist an identity (or very substantial overlap) of their *reference*, sufficient to underpin the continuity and objectivity of scientific progress. We can dub this view *the stability of reference thesis*; among its notable adherents should be counted Israel Scheffler and Hilary Putnam.[5]

In its customary, strong form, the stability of reference thesis is a nonstarter, however. It should, to maintain the appropriate measure of 'objectivity' and 'continuity' in science, construe 'reference' in an informal guise. The reference of a scientific term may then, on this view, be regarded as externally and theory-independently determined, and held stable throughout changes of theory. One might try to augment this view by means of some ontological theory of the metaphysical realist sort. But this would be tantamount to buying objectivity at the expense of never being able to specify exactly *how* reference can be preserved under changes in use and belief. Reference may be objectively fixed, but one can never have direct empirical access to it. If, on the other hand, one considers the epistemological question of what, as a matter of *empirical* fact, is the reference of a given scientific term, one can only be guided by the best available theories and by current usage. But this seems to mean giving up any general account of how reference can be impervious to changes of belief and theory. The driving force of the incommensurability thesis is, after all, directed at the epistemic issue of how rational are our *choices* of theory, how objective our beliefs. Even if one were to claim that there is an objective fact of the matter as to what and how scientific terms refer, one's *knowledge* of reference can only be conjectural, and, like scientific theories, subject to change. The process of testing and comparing rival theories, likewise, is subordinate to 'facts' and hypotheses which are at bottom fallible; and here the ideal of objective, 'external' reference is scant consolation.

The various attempts to embellish the stability of reference thesis with philosophical theories of meaning seem to hold out little hope for rescuing it from the dilemma I have just outlined. Putnam, for instance, has tried grafting Kripke's account of proper names (the so-called *causal* theory of reference) onto the semantics of scientific terms generally. By divorcing reference from meaning or Fregean sense, the causal theory might indeed explain referential stability in the face of changes in belief and meaning. But there are severe obstacles in its path. When pressed to spell out exactly what the referent of a general term is, the causal theory seems forced in the end to appeal to properties that the concept named by the term possesses *in virtue of* a given scientific theory.[6] If this observation is correct, the idea of referential stability is once again impotent, because causally determined reference is powerless as a practical aid in establishing cognitive intertheory relations, for example by constructing translations which match up the co-referring expressions of two theories.

A more discerning defence of referential stability from the standpoint of scientific realism has been offered by Jarret Leplin (1979). Leplin criticises both Scheffler's approach, which he identifies with the Fregean view of meaning and reference, as well as those accounts of the Kripke—Putnam type based on a non-Fregean, causal theory of reference. In particular, Leplin musters detailed historical evidence — concerning especially the reference of 'electron' in physics — in order to show that neither approach guarantees the right sort of referential continuity to block the inference from meaning variance to historical relativism. Having rejected the standard account, he provides a new one of his own.

His suggestion, in brief, is that, independent of the original manner of introducing a scientific term t, there are certain circumstances in which the reference of t can be seen to be preserved through successive, alternative theories in which it occurs. A case in point would be where two such theories T and T' employing t agree on a certain prediction E, and in each theory the purported reference of t is a necessary condition for E; so that the truth of E justifies an existential commitment to whatever t names. As rivals, T and T' will not concur on all matters related to E. In fact, Leplin takes as typical the case that (the later theory) T' rejects (the earlier theory) T's *explanation* of E. Thus, T' may ascribe different *properties* to t, yet still retain for this term the same reference that it had in T. And, according to Leplin, the evidence for this last claim is strong, even if one has no direct access to the actual referent of t in T.

Leplin's argument is intended to undermine historical relativism and support scientific realism without rejecting the thesis of meaning variance; in fact, meanings are allowed to change in the transition from T to T'. But the argument seems too weak to achieve this. First, the assumptions imposed are severe: in particular, the requirement that T and T' share a common prediction for which the existence of t is essential. Hence it is unclear whether they would apply in the majority of cases where, for instance, incommensurability has been claimed. Secondly, even if the premises were typically satisfied, I cannot agree with Leplin's conclusion and the inference he draws from it. E can be imagined as the statement of a certain result or fact. But one has no guarantee that the meaning of E is the same in both theories. If T and T' are genuinely incommensurable in the sense that no translation between them is possible, then it is merely an accident of syntax that this statement can be derived from each theory.

Conversely, in order to make a legitimate comparison of the explanations of E offered by the two theories — and thereby provide grounds for co-reference — one needs to know that sentences involving t in the language of T can be translated into the language of T'. Given *this*, and knowledge of how the translation proceeds, one might well obtain suitable evidence of referential stability. But plainly an argument which purports to counter historical relativism cannot take the assumption of translatability and commensurability for granted; though this, in effect, is what Leplin's claim does.

Perhaps the real difficulty here lies in the notion of 'reference' itself. Perhaps referential semantics is simply the wrong way to go about resolving the paradoxes of meaning variance. Moreover, a strong suspicion emerges that the error, if there is one, in the incommensurabilist's claim must be rooted in the *holistic* theory of meaning to which he seems committed. (Witness the radical form of holism that Kuhn evinces in the quotation above.) Is there, then, a suitable molecular account of meaning, unattached to the notion of reference, that would respond adequately to Kuhn? One obvious candidate is the 'anti-realist' semantical theory developed by Dummett, Prawitz, Tennant and others.[7] It is based on the idea that meaning is determined by *use* as manifest in observable linguistic behaviour, but it is molecular in supposing that "the meaning of any well-formed expression in our language depends in a rule-governed way on the meanings of its constituent simple expressions" (Tennant, 1987). What is more, since meaning is held to be controlled not by the totality of use, but rather by some *central features* of use, this approach could yield some philosophical underpinning to the idea that constitutive meaning could be invariant under global changes in belief and linguistic practice.

This theory has achieved some recognition (and notoriety) of late, but only in a rather restricted (if central) domain: it purports to provide an adequate proof-theoretic semantics for the logical operators (under intuitionist or minimal relevantist readings). It would seem to require a Herculean effort, however, to transpose the theory successfully into the realm of scientific concepts at large. What would correspond in this more general setting to the notion of canonical proof in intuitionistic mathematics?[8] How could one identify in the framework of mechanics, say, exactly those features of use that are constitutive of meaning? Problems of the latter kind do not seem to have been squarely tackled, even by philosophers, like Sellars and others, who commend an explicit appeal to the *role* and *function* of concepts as determinants of meaning, in order to rebuff the conceptual relativism of Kuhn and Feyerabend. For instance, Gutting (1973) remarks that there are numerous similarities in the rules of usage governing classical and relativistic 'mass'. Quite so. But he gives no argument for why the similarities he has in mind are precisely those that should control the meaning of 'mass' in the two theories. Neither does he offer a general account to explain why as a rule the similarities rather than the differences in usage of a given term are to be relied upon for inferring continuity of meaning across changes of theory, rather than discontinuity.

In the last few pages I have been discussing meaning and reference as construed in the *informal* sense; that is to say, on the understanding that a scientific theory is not logically reconstructed, and that the reference of a term is, on a realist interpretation, identified with the set of 'real world' objects to which it applies. On this view, however, a scientific theory is a loosely defined entity, and there is ample scope for terminological and substantive disagreements about how the theory should be formulated and what are the intended meanings and referents of its central concepts.

An obvious gain in clarity is afforded by the perspective of *formal* semantics, wherein an empirical theory is treated as a logically reconstructed entity, and questions of sense and reference are dealt with in a formal (e.g. model-theoretic) framework. The chief advantage is that the extension and intension of a scientific term may be precisely determined, say along Tarski, Carnap, or Montague lines. Naturally, there is a price to pay. In this case 'reference' loses any pretensions it may have had to being an 'absolute' notion, by becoming relativised to the formal framework. The are at least two degrees of freedom or relativity of formal reference. The first is that reference and other semantic values are subordinate to the kind of semantic interpretation laid down by the metatheory, e.g. what kind of 'models' are selected; they could be two-valued or many-valued, Kripke or Tarski models, and so on. Secondly, even when it has been settled what is to count as a *bona fide* semantic structure, an empirical theory does not generally determine a unique, intended model, but rather a *family* of models. The extension of a term is thus fixed for each model (of the theory, or of some otherwise distinguished class), but not fixed absolutely.

Relativity of reference is a basic fact of life in formal semantics, and a fact to which logicians have long grown accustomed. Curiously, it has lately been the cause of consternation and controversy in the philosophy of language, sparked off by Putnam's allegations that the underdetermination of (formal) reference by theory undermines the cogency of *realism* as a philosophical thesis.[9] Putnam has been soundly admonished for toying with such wild ideas. But few commentators seem to have locked in on the real error of Putnam's ways, which consists in his super-imposing properties of formal reference onto the setting of informal semantics in which realism is usually discussed. Putnam just slides casually between these two areas of discourse, formal and informal, as if arguments drawn from one area would apply without change to the other.[10]

Once reference goes relative, the simple formula "translation is a reference-preserving mapping of expressions" requires a modicum of re-evaluation. The notion "sameness of reference" may seem clear enough when applied to straightforward terms in ordinary language, e.g. colour words in English and German. In most contexts 'red' is co-referential with 'rot', and may be translated by it; a single model of (the appropriate fragment of)

English is paired with a suitable model of (the corresponding fragment of) German. Here, subject to context (which can often be restricted or controlled anyway), the source and target languages have each a single interpretation by means of a model that specifies determinate extensions (glossing over the problem of vague predicates). Moreover the two models share a common universe of discourse in which co-extensionality is decided. This state of affairs — which is, of course, highly idealised, even for limited fragments of natural language — does not apply even as a first approximation to formalised scientific theories. Even if one can come up with a suitable pairing of models of the source language with those of the target language, one has no reason to expect that corresponding models will have identical universes. This was the problem that we encountered already in Chapter 2 when criticising Balzer's constraints on proper translation (especially (2.0)(iii)) and when trying to make more explicit Kamp's criteria of adequacy for translation.

This fact presents no intrinsic barrier to translatability, however. It serves only as a reminder that in formal semantics co-extensionality is, first and foremost, a concept that is relativised to a given model or models; secondly, that it is anyway not *quite* the concept that one is searching for in order fully to grasp the process of translation. Fortunately, one can translate, perfectly correctly, in many cases where sameness of reference in the naive sense is not the appropriate criterion of adequacy. One can translate geometric predicates into the language of arithmetic, concepts from rigid-body mechanics into the language of particle mechanics, and intuitionist logical operators into the framework of classical logic.[11] In the first two cases one simply deals with pairs of models that have different domains (as well as relations); in the last case the concept of 'model' has altered too, and one has to relate classical structures with Kripke models.[12]

From these and other examples of successful translations, one can obtain a fairly precise characterisation of what it may mean in general to say that 'semantic values' are preserved.

3. INDETERMINACY OF REFERENCE

The relativity of formal reference in no way implies, therefore, that formal semantics cannot be used to account for conceptual continuity through changes in belief and theory. But we need to consider in rather more detail what sort of account might be given, and what kind of continuity it might guarantee.

Two particularly interesting approaches to this problem have been developed, by Hartry Field (1973), and by Marian Przełęcki (1979; 1980; see also Przełęcki, 1969; 1976). Though independently worked out, there is a certain resemblance between these two accounts, and the conclusions they reach. On the whole, Przełęcki's is logically more sophisticated; Field's pays closer attention to historical examples, which perhaps explains the greater impact it has had on philosophical debates about meaning variance and incommensurability.

Both Field and Przełęcki adopt the view, which I earlier associated with logical empiricism, that certain terms of a supplanted scientific theory may become referentially indeterminate from the standpoint of a later theory. The situation of 'mass' in classical mechanics would be a typical case for both authors; relativity theory being, of course, the later standpoint. Rather than conclude, however, that words like Newtonian 'mass' should be regarded as completely uninterpreted, Field and Przełęcki set out to explain *how* sentences containing this term might be said to have determinate truth values. Put more generally, then, their problem is to give a semantical account of how a sentence can be determinately true-or-false even though it contains terms or constituent expressions that lack definite denotation.

Field argues that this very problem situation creates a serious difficulty for standard (Carnapian or Tarskian) semantic theory (which he calls *referential* semantics), and requires for its solution a major revision of the theory. His reasoning seems to be that, according to referential semantics, the truth values of sentences are uniformly determined by the extensions of their component predicates and names. But if the latter possess no determinate extensions, how can it make sense to inquire after the truth value of the whole compound expression? As a critique of semantic theory, this argument looks unconvincing, however. In fact, the theory is intended to apply in one of two situations: (i) a universe of discourse for a language L is given, and the language (vocabulary) is assigned a single interpretation which fixes determinate extensions over that universe. Or (ii) it is specified what is to count as a *model* $\mathfrak{M}$ (structure, semi-model, realisation) for L, where $\mathfrak{M}$ likewise assigns unique extensions to each symbol of L. In both cases, the truth value of a sentence is defined recursively on its simple constituents; either truth-in-L is defined, as in (i), or truth-in-$\mathfrak{M}$ is defined, as in (ii). There is no question here of standard semantic theory being inadequate, as Field seems to think, because the theory simply *defines* truth values on the basis of *determinate* reference.

Perhaps Field only wishes to point out that, in the case of scientific languages, (i) does not generally apply, due to referential indeterminacy. But this is hardly a criticism of Tarskian semantics. For, of course, no one would *want* to employ (i), except where L can be assumed to be semantically unambiguous, since (ii) is always available and can be used to cover the case that L has vague or indeterminate predicates, or should for any other reason be assigned multiple interpretations. This is precisely the strategy adopted by Przełęcki, who argues that all empirical languages display ambiguities and indeterminacies, due to the inevitable vagueness of qualitative terms and the imprecisions of measurement associated with quantitative terms. He proposes, therefore, that the intended interpretation of a scientific theory be specified by a family M of models. A sentence φ in the language of the theory is then *true* (*false*) if it is true (false) in all models in M; otherwise it may be said to be *indeterminate*.[13]

Field's argument for referential indeterminacy takes a different route. He claims that it makes no sense to ask for the *real* reference of Newtonian 'mass', because from the standpoint of special relativity, RM, this term could just as well refer to *rest mass* as it could to *relativistic mass*. Since there are no grounds (and there never will be any) for preferring one of these two interpretations over the other, there is simply no fact of the matter as to what Newtonian 'mass' refers. Rather than rest content with the correct but uninformative conclusion that Newtonian 'mass' does not *fully* denote either of these two concepts from RM, Field proposes that one thinks of this term as *partially* denoting each of the relativistic concepts. To show how partially referring expressions might combine to form sentences with determinate truth values, Field uses the same method as Przełęcki: relativising truth to a given model. He then calls a sentence φ *true* (*false*) if it comes out true (false) in every model that assigns to the constituent terms of φ relations (or individuals) that they partially denote.

Field maintains that the notion of partial reference thus renders denotationally comparable terms like 'mass' as they appear in CM and RM. Moreover, some sentences in the language of CM will turn out to be determinately true-or-false from the relativistic perspective. It is evidently crucial for the success of this analysis that one is actually able to pick out the partial referents of a defunct scientific term and effectively determine what sort of denotational *refinement* (to use Field's expression) it undergoes in the transition to a new theory. Unfortunately, Field does not supply any guidelines for how this could be achieved in general; and his approach is also open to the criticism that it cannot assure us that most or even any sentences of a supplanted theory will enjoy, from some later standpoint, determinate truth values. The major difficulty with his view is that it seems to suggest that all scientific terms, current as well as superseded, are liable to be referentially indeterminatc. One can well imagine, therefore, that some future theory will show even the partial referents of Newtonian 'mass' (as they are currently reckoned) to be them-selves only 'indeterminate' quantities.[14] This seems to make nonsense out of Field's claim that his method allows one to assign (today) actual, determinate truth values to certain sentences of Newtonian mechanics.

Przełęcki, on the other hand, argues not for the sameness of *partial* reference, but for the *identity* of *full* reference in *some* intended models of two theories under comparison. In the sim-plest possible case, if M and M' are the respective families of intended models, and t is a given term belonging to each theory, the claim is that for some models $\mathfrak{M}$ in M and $\mathfrak{M}'$ in M' it will turn out that $\mathfrak{M}(t) = \mathfrak{M}'(t)$.[15] One of the advantages of Przełęcki's approach is that it is based on a definite meaning theory for scientific languages. The classes of intended models which pre-scribe the actual referents of a given term are determined accord-ing to the methods of *semantic empiricism*, formulated along Carnapian lines. Moreover, his general claim about overlap of ref-erence is plausible (modulo the stated assumptions) in those cases like CM and RM where one theory corresponds in the limit to the other; even though the details have not been worked out in full.

The chief drawback that I detect in Field's and Przełęcki's proposals is not that they cannot supply some measure of conceptual continuity in science, but that the amount of conceptual continuity falls short of what one requests and requires of rational theory appraisal and choice. The problem is especially acute in Przełęcki's case, because the meaning theory he adopts in order lend weight to the thesis of referential comparability (and which gives him the upper hand over Field) effectively rules out any possibility of (meaning-preserving) intertheory translation.[16] One might almost be inclined to say that, for him, the overlap of reference is bought at the cost of untranslatability. Yet it is worth emphasising the point once again that translation of some sort is absolutely indispensible for achieving a cognitive comparison of rival theories. The reason is simply that even if the theories in question could be shown to deal with co-referring or partially co-referring concepts, it would not follow from this that their respective *claims* would be commensurable and could therefore be rationally appraised. In fact, Przełęcki's line of approach provides an excellent example of how referential comparability and incommensurability might arise together.

For Field, the situation is somewhat different, though he too concedes that there are untranslatable sentences of classical mechanics: an example would be the statement equating momentum with the product of mass and velocity. But the grounds for nontranslatability are, in Field's case, not very clearly presented. Having initially rejected the idea that Newtonian 'mass' can be identified with either of the terms of the same name in RM, it *seems* as if Field infers the untranslatability of (all occurrences of) 'mass' on the basis that any adequate translation *would have to be* literal in the sense of transforming 'mass' into either 'rest mass' or 'relativistic mass'. This would certainly be a puzzling claim in view of his assertions about referential indeterminacy, since these would tend to point us in the direction of nonliteral translations.

Field's claims to have resolved some of the difficulties surrounding the problem of incommensurability are somewhat optimistic in view of what he *actually* achieves. For, as he openly confesses:

> Newtonian physics *as a whole* is objectively false, but there is no fact of the matter as to how the falsity of the theory as a whole is to be distributed among the individual sentences of the theory.
>
> (Field, 1973, p. 474)

This is copybook Duhem—Quine, and Field intends it as such. But the point to remember is that in comparing CM with RM (or any two rival theories, generally), one is not simply confronting one theory with conflicting evidence or test statements. If one *were*, then by Duhem's argument one would indeed be unable to redistribute falsity (and truth) over sentences of the theory individually. when comparing CM with RM (or any subsequent theory available) one is attempting to explain with the help of the later theory *why* and *where* CM went wrong, and, in particular, where it was at least approximately correct. This means that one's goal really *is* to distribute truth and falsity (or approximate accuracy, or whatever) across the sentences of the refuted theory.[17] Although Field actually qualifies the last clause of the quoted sentence — presumably because on his analysis *some* sentences of CM do have determinate truth values — he does not, in my view, go far enough. For one thing because, by the objection I raised earlier, it is unclear how many statements of classical physics remain, for him, factually meaningful; for another, because one would like to know of the truth-valueless or indeterminate sentences to what extent they are approximately correct.[18]

The chief problem, then, with the Field—Przełęcki type of analysis is that it seems to fail of translatability and thus the kind of cognitive comparability that one is aiming for. There are, however, philosophers, like Mary Hesse (1977), who have drawn different lessons from the partial reference approach. According

to Hesse, Field is at bottom demonstrating that terms such as
'mass' in CM *ambiguously* refer, sometimes to rest mass, sometimes to relativistic mass, depending on context.[19] Therefore, she
holds, one can and should *translate* Newtonian theory into the
language of RM, employing a *principle of charity*, and relying on
context to see which of the two relativistic terms one should
correlate with 'mass'; sometimes the one is appropriate, sometimes the other. Hesse herself does not try to carry out this
translation, which is just as well because the problems involved
are formidable.[20]

We can illustrate the difficulty by attending to one such attempt
at translation which fails dramatically. Michael Levin begins his
(1979) paper by observing that 'mass' means something different
in CM than in RM, and he remarks perspicuously that this would
suggest we try our hand at a nonliteral translation. He then
proceeds to try to define precisely what he has just ruled out: a
literal translation of CM into RM (or better: modern English
permeated by relativistic expressions). Letting m stand for 'Newtonian mass', m_0 for 'rest mass' and m_r for 'relativistic mass' (i.e.
$m_0/\sqrt{(1 - v^2/c^2)}$), Levin proposes the following 'translation'
(which for obvious reasons I cannot paraphrase);

> Let $F(m)$ be all the things Newton said about m that are true of m_0,
> and $F'(m)$ all the things Newton said about m which are true of m_r.
> We can simply let m divide when we translate. If M is an otherwise
> acceptable mapping from Newtonian English into ours, we let $M(m)$
> $= m_0$ if m occurs in F, and let $M(m) = m_r$ if m occurs in F'.
>
> (Levin, 1979, p. 410)

This 'translation' is both formally and methodologically awry.
First, one can suppose that M is intended to map *sentences*, but
in that case it is unclear what the expression '$M(m)$' is to mean.
Presumably it can be taken to mean something like "the result of
translating a given sentence φ which contains 'm'". Secondly, by
this method much of what Newton said about mass will be

untranslatable; for example, (i) any sentence that cannot be made to come out relativistically valid by replacing m by either m_0 or m_r, and (ii) any conjunctions of sentences whose individual conjuncts are relativistic valid under *different* substitutions for m; similarly for other combinations of sentential connectives. Thirdly, all sentences which can be translated (i.e. all 'values' of M) are relativistically *true*. Fourthly, the assumptions that M could be "an otherwise acceptable mapping of Newtonian English into ours" is, to say the least, questionable. In order for the domain of M to contain anything interesting that Newton said about mass, it would have to include sentences involving 'force', 'velocity', and so forth. But this would imply that there is already an acceptable translation of *these* terms, quite independent of what Newton said about mass (otherwise the expression '$M(m)$' would have no precise meaning). This is quite implausible, in view of the fact that Newton shaped his beliefs about all the concepts of mechanics *together* in the context of a single theory, CM. Fifthly, this 'translation' could never assist one in understanding what Newton 'really meant' (and thus evaluate how correct he was). It can only be a mapping of sentences to sentences guided by the purely expedient rule: "interpret Newton as being right, by the lights of RM (or modern physics), or else don't interpret him at all." The mapping M cannot be a way to re-express accurately in a later vocabulary what Newton actually said, because it is *defined using the notion of truth in RM* (or in current theory). It therefore completely begs the question as to whether Newton spoke the truth or not.[21]

This last objection is perhaps the most serious one, and calls for some comment. Levin correctly observes that translations between natural languages are not as a rule word-for-word, neither do we unfailingly translate a given source language word by the same target language expression; we often vary the translation according to context. He infers from this that consequently one should not be alarmed if the term '*m*' from CM does not correspond to a *single* word from RM, and if occurrences of '*m*' may sometimes be mapped onto *different* expressions of RM. This is unobjectionable, but the point is not well taken. Translation is essentially a rule-governed process, and the operative rule may well require that an ambiguous term is 'divided' in translation. But the criteria for applying the division must be based on features of the *source language*, and thus in principle expressible in the source language itself (perhaps with some metalinguistic terms thrown in). Levin's translation 'rule', in contrast, is formulated *in terms of* features of the *target language*, RM, or more accurately a hypothetical metalinguistic truth predicate for that language or theory.[22] The chief cognitive aim of the enterprise is thus thwarted right from the start. That aim is to provide in terms of CM (plus appropriate metatheoretic apparatus) a rule whereby the sentences of its language are translated into the framework of RM in such a way that their truth value (accuracy, degree of confirmation, and so forth) can be simultaneously or subsequently assessed on the basis of RM. In the present context this is what it means rationally to appraise one theory in the light of another. Clearly, Levin's procedure cannot meet this aim because his rule of translation begs the question it was designed to answer and preempts the issue of rationality.

4. KUHN AND FEYERABEND AGAINST TRANSLATION

So far the picture looks anything but rosy for the rationalist and commensurabilist. The path to cognitive theory comparison by way of referential stability looks steep, and the rewards, in the end, unappetising. And, if my last argument is correct, the method of translation will have to be rather more sophisticated than some authors imagine, if it is to succeed at all. Could it be that an adequate translation is simply not available for the cases like classical and relativistic mechanics which most interest us? Or is translation possible, but merely difficult? Are there any solid empirical or philosophical grounds for claiming that nontranslatability is really the case? Kuhn and Feyerabend have each tried to answer the last question affirmatively. A brief examination of some early statements by Feyerabend and some recent arguments by Kuhn addressing this issue will help to place my own approach to translation in the right perspective.

Kuhn, in his (1983), quite properly distinguishes the process of language acquisition and interpretation from the process of translation. The historian, for Kuhn, is first and foremost an *interpreter* of texts, an *acquirer* of a language that may be very different from his own. He may, though he certainly need not, resort to *translation* into current idiom, in order to interpret, understand and evaluate ancient texts and outmoded theories. Kuhn's historiography is by no means uncontroversial. Already at this point one might be inclined to ask how much of the process of interpretation involves or presupposes translation. And one could surely make out a strong case for favouring historical reconstructions that are informed and influenced by the subsequent developments in a scientific discipline. I shall not dwell here on Kuhn's questionable historiography, but I do want to suggest that his separation of the two activities of interpretation and translation achieves little in the way of backing the argument for untranslatability.

What Kuhn sets out to show, in the first place, is that it is not inconsistent to maintain that a historian can interpret and understand a theory that has been superseded, even if that theory cannot be fully and adequately translated into the language of the theory that supplants it (or into current language). One can learn to understand eighteenth-century phlogiston chemistry and Newtonian mechanics, for instance, although both theories contain untranslatable terms. This is Kuhn's rejoinder to many of his critics who have accused him of adopting double standards, one for the historian, one for the scientist and philosopher. Kuhn then tries to convert this point into a positive argument for the untranslatability of CM, by appealing to the *learning* process.

> In learning Newtonian mechanics, the terms 'mass' and 'force' must be acquired together, and Newton's Second Law must play a role in their acquisition. One cannot, that is, learn 'mass' and 'force' independently and then empirically discover that mass equals force times accelera-tion. . . . Though 'force', say, may be a primitive in some particular formalization of mechanics, one cannot learn to recognise forces without simultaneously learning to pick out masses and without recourse to the Second Law. This is why Newtonian 'force' and 'mass' are not translatable into the language of a physical theory (Aristotelian or Einsteinian, for example) in which Newton's version of the Second Law does not apply. To learn any one of these three ways of doing mechanics, the interrelated terms in some local part of the web of language must be learned or relearned together and then laid down on nature whole. They cannot simply be rendered indi-vidually by translation.   (Kuhn, 1983)

I do not want to contest this claim as a statement of how Newtonian mechanics is *learnt*, presumably, for Kuhn, through exemplars: paradigmatic applications and problem solutions. I also agree with Kuhn that terms like 'mass' and 'force' in Newtonian physics acquire their meaning *together*, in the context of a theory like CM. (This was one of my objections against Levin, just now.) However, it is far from clear why the factors the Kuhn cites as relevant to the *learning* of CM should at the same time make the *translation* of that theory impossible. Moreover, even as far as interpreting and understanding CM today is concerned, one cannot simply *forget* the later developments in mechanics. One can learn to apply classical masses and forces to solve mechanical problems, *as if* CM were true, or approximately correct within acceptable limits. But if one also knows or believes that CM as a whole is *false*, or as a matter of fact inadequate for solving certain problems, this knowledge is bound to affect one's understanding of CM in some way. If relativity theory, RM, is the point of reference, one will plainly accept certain limitations on the adequacy of classical concepts and problem solutions formulated in CM. By the same token, RM also provides an insight into *why* CM could have been thought true in the past, and in what circumstances it may still be retained for certain applications in the future.

Far from entailing that CM cannot be translated into the language of RM, these observations suggest that it ought to be translatable. Of course, one would not normally maintain that CM is *learnt* by first translating it into the language of relativistic physics. Similarly, one *interprets* Newton's *Principia* principally by attending to what Newton wrote there, not by first turning to a translation of it into the language of modern physics. However, it seems to me that the historian *qua* interpreter and the philosopher *qua* rational reconstructor have compatible and complementary tasks to fulfil; their aims are closer than Kuhn seems to imagine. Scrutinising the passage from classical to relativistic mechanics, the historian explaining this process cannot interpret CM as true (or empirically adequate) and still provide a rational account of how RM was developed and came to be better confirmed. He considers, rather, the various anomalies and conceptual problems that CM faced, and the experimental findings which, interpreted in the language of CM, indicated empirical inadequacies in that theory, yet, reinterpreted in the language of RM, became *its* problem successes.

The translator, or rational reconstructor, has broadly similar goals here. A translation of CM into RM under which Newtonian laws came out 'true' or relativistically derivable would evidently be inadequate for explaining rational choices. A literal or homophonic translation, on the other hand, would reproduce the syntactical inconsistency between the two theories, but would fail to bring out the conceptual disparities and differences of meaning in their central terms. The way we *understand* CM, however, is not simply as false, but as *approximately* correct in some situations, thoroughly inaccurate in others. This understanding can and should be reflected in translating CM into RM. The translation should bring out the fact that some classical equations are relativistically derivable in the limit, and thus clarify the sense in which RM improves on CM, and CM is inadequate *in the light of* RM.

If this means one first has to learn RM in order correctly (or better) to interpret CM, I have no objection. Rationality would cease to be critical if it demanded that a long-lived theory like classical mechanics possessed a *unique* intended interpretation, fixed forever. Even in its own lifetime, classical mechanics underwent revisions and reinterpretations through the work of Hamilton, Mach, Poincaré and many others. I see nothing wrong in agreeing with Kuhn that theory change might admit a fully rational reconstruction only *with hindsight*, that theories like CM could not be *refuted* by empirical findings alone, that so-called decisive experiments have to be reinterpreted according to the framework or theory. None of this undermines the possibility of translation, even of the kind I recommended. On the contrary, the test of a good translation is how well it reproduces these ideas.

Perhaps Kuhn only wants to argue the point I already conceded, that one's understanding of Newton's theory *as if* it were true or adequate is bound to differ from one's understanding of Newton's theory as false or only approximately adequate. But if this were the force of the incommensurability thesis, it would have little or no harmful effect on the empiricist conception of rationality. Fully to understand CM *is* to understand it in the light of later developments in mechanics and to explain those rationally. To insist, say, that Newton's interpretation of CM is the only admissible one might be tantamount to demanding that CM *can* only be correctly understood when understood as correct.[23] In that case, an adequate translation would indeed, to preserve meaning, also have to preserve the 'truth' of the theory, and here one would run foul of our earlier strictures on rationality. But this is just a variant of the more general claim that one can never translate false beliefs, and this I reject as fallacious. In translating from CM into RM, what I think one should be looking for is not to preserve 'truth', nor syntactical identities, but to preserve something that might be very loosely termed 'truth-in-the-limit'; for this seems to be the sense in which 'truth' (empirical adequacy, or whatever) really does transfer from CM to RM.

This brings me to what is perhaps the major area of conflict between Kuhn and his critics. Most disputants in the rationality debate accept that, in the case of classical and relativistic mechanics, for instance, there exists a precise logical relation between RM and a reformulated version of CM which is entailed by RM in the limit of low particle velocities. For many of Kuhn's critics, this is sufficient for having continuity, rationality, the explanation of one theory by its successor, and so on. But Kuhn sees this as problematic, because he cannot accept that the reformulated theory is really CM; it is, rather, CM suitably corrected and approximated. And the fact that *this* theory corresponds, or is otherwise logically related, to RM does not make CM commensurate with RM. Kuhn has I believe a valid point here, but makes an unsound inference from it. My suggestion is that the disagreement between Kuhn and his critics could be resolved by adjusting the initial claim to this: that RM bears an exact logical relation to a *translation* of CM into relativistic language. This should be acceptable to the critics, most of whom never denied translatability anyway. It also *ought* to be acceptable to Kuhn, because there can be no 'discrepancy' between CM and its translation; they are in effect the 'same' theory in different languages. He could, of course, deny that the translation is an adequate one, but it is hard to see what grounds could underpin such a claim, particularly since, as I hope to show in the next chapter, the translation I have in mind achieves most of the things that could be desired of it, including much that Kuhn has doubted a translation could achieve.

Kuhn offers some very general remarks on criteria of adequacy for acceptable translations, but none is precise enough, it seems to me, to 'translate' into specific conditions which one could apply to the case CM/RM at hand.[24] Feyerabend (1962), however, does supply a rather tighter argument for nontranslatability, and from this I think one might extract a more or less definite set of adequacy conditions.

At the core of Feyerabend's argument for incommensurability is a radical critique of the notion of theory replacement by theory reduction. He attacks what he takes to be the standard empiricist view, whereby, in a case of theory change, the old theory is *reduced* to its successor and is deductively explained by it. According to Feyerabend, this standard view of reduction comprises three essential premises:

(i)   that the theories involved are mutually consistent;
(ii) that there is no change of meaning among their central terms;
(iii) that the laws of the earlier theory are logically derivable, with the help of auxiliary assumptions, from the laws of the later theory.

Evidently, in the case of classical and relativistic mechanics, the first premise does not hold. Feyerabend's arguments against (ii) are well known and need not be repeated here; some of them we have already rehearsed anyway. And it seems as if (iii) is almost bound to fail if (i) and (ii) do not stand up. One of the principal reasons for rejecting (iii) in this case is that there is not even a restricted 'domain' in which the two theories are in genuine agreement.[25] For, the limiting conditions under which some laws of

RM 'pass into' classical counterparts are actually counterfactual, or inconsistent with RM itself. It is easy enough, therefore, to express a purely mathematical relation between the equations of CM and RM, but no *conceptual* relations between their terms are forthcoming, nor is there a meaning-preserving translation.

According to Feyerabend, therefore, RM does not explain the valid part of CM, at least not in any traditional sense of deductive explanation; and the absence of any translation precludes the usual cognitive comparison of the theories, for example a comparison of their empirical contents or their degrees of truthlikeness. Feyerabend himself, of course, is only too willing to give up the deductive concept of explanation here, and many philosophers have subsequently followed his advice. To fit such cases, 'structural' and 'counterfactual' models of explanation have been proposed which depart in various respects from the deductive ideal. On the other hand, models of so-called *approximative* explanation, which purport to trade exact for approximate entailment in (iii) with little or no epistemological 'loss', tend in my view to raise more questions than they solve.[26]

Should one even try in the case of CM and RM to preserve something like a reductive relation? Or should one go along with Feyerabend's rhetoric:

> Why should the relativist [relativity theorist] be concerned with the fate of classical mechanics except as part of a historical excercise? There is only one task we can legitimately demand of a theory and that is that it should give us a correct account of the *world*. What have the principles of explanation to do with this demand?
> (Feyerabend, 1970, p. 227)

The view underlying Feyerabend's statement here is, I believe, historically inaccurate and methodologically unsound. It is questionable historically because as a matter of fact we believe that theories are not born into a cognitive vacuum, but that they arise in response to problem situations and they spawn in their turn new problems, empirical and conceptual, which new theories are then developed to resolve. In this sense one can say that relativistic mechanics was a development out of classical mechanics *in response to* anomalies and conceptual problems which that theory faced. It was, moreover, an important part of the *success* of RM that it could *explain* the limitations of the Newtonian world view, besides furnishing its own, more adequate view of mechanical phenomena. This, I think, shows the weakness of Feyerabend's professed 'radical' methodology: it deprives scientific theories of an important part of their explanatory and heuristic power, and their problem-solving ability.

If one grants that it is desirable that RM should be seen to explain the successes and failures of CM, then it will be natural to search for a reductive type of relation connecting them, one that is both syntactically and semantically grounded. It will have to include translation, of course; and by our previous arguments the translation will not be a literal embedding of CM into RM. In line with this, one will also have to give up the premise (ii) of meaning invariance. But I doubt that this should anyway be taken as an intrinsic feature of any derivational model of explanation and reduction. What seems to be important in this context is that conceptual disparities and differences of meaning can be suitably *corrected*, in other words that a nonliteral translation is feasible. The translation should, of course, meet the usual semantic requirement of preserving 'reference' in an appropriate way; and it should, hopefully, comply with syntactic constraints like being recursively definable within apparatus available to us in the source language (and metalanguage) itself.

On balance, I think one can single out four essential formal and methodological properties that a translation should possess in order for it to respond adequately to Feyerabend's challenge. First, it should render all sentences of CM in the language of RM. Secondly, it should preserve the basic incompatibility of CM and

RM in the sense that not all classically correct claims will be relativistically valid under translation. But, thirdly, the translations of Newtonian laws should be derivable from RM in the limit. And, fourthly, a deductive explanation of CM from RM should be thereby achieved, and the 'successful part' of CM should be absorbed by RM under translation. In the next chapter I shall indicate how a translation meeting these requirements can be formulated.

7

TWO ROUTES TO COMMENSURABILITY

1. COMPARABILITY, RATIONALITY, TRANSLATABILITY

Up to this point I have examined several of the more prominent directions philosophers have explored in order to try to establish the rational comparability of scientific theories. The results have not always been encouraging. As we saw, Stegmüller maintains that theories can be cognitively compared when viewed as *structures*, though they may be noncomparable as sets of statements. The fallacy here, I argued, lies in the belief that the structural and linguistic perspectives are so radically different that one of them may permit rational comparisons where the other does not. I showed in Chapter 1 that, under a sufficiently liberal interpretation of 'logic', 'language', 'deduction' and so forth, structurally comparable theories are also logically relatable in the syntactic and semantic sense. Stegmüller's claim to have dissolved the problem of incommensurability and irrationality by the strategy of switching to a new metascientific paradigm thus collapses; whilst his and Balzer's attempts to reinstate the claim by proffering strict criteria for commensurability must also be rejected, on the grounds argued in Chapter 2.

Laudan's idea, on the other hand, was that rival theories can be compared for their cognitive contents, providing that 'content' is construed not in terms of truth and truthlikeness but in terms of PSE, or problem-solving effectiveness. My reply to this, in Chapter 3, was that comparisons of PSE are no consolation for the rationalist unless the theories of rival research traditions can be conceptually related in such a way that shared empirical

184

problems and methods of evaluation are forthcoming. This, I believe, shows that if PSE is to work as an index of rational progress, research traditions must be in Laudan's sense commensurable, and translation should be available to indicate how problem solutions and anomalies can be transferred between traditions. In Chapters 4 and 5 I sketched a framework in which these issues can, at least as a first approximation, be treated more precisely.

My claim against Laudan — that intertheory translation is a necessary condition for establishing rational comparability — also holds good, I contend, for any other measure of cognitive content that one may care to insert in place of PSE. Comparative evaluations of theories with respect to predictive success, explanatory and heuristic power, degrees of confirmation and verisimilitude, and so on, likewise require translatability of a suitable kind if they are to function as cognitive criteria of rational appraisal and choice. This is one of the primary reasons why I found occasion in the last chapter to question the stability of reference approach and to doubt the utility of even the more plausible semantic theories that make allowances for referential indeterminacy.

There might be an objection here that the requirement of translatability is unnecessarily severe. I dissent. I think a good case can be made that any other reasonable candidate will turn out to be at least as strong. Take as an example the popular idea that to compare two rival theories it suffices to find a language rich enough to express each of their concepts and claims. Suppose, for instance, one is given such a language L that is a common extension of the languages of two theories under comparison. Now, if the theories concerned are conceptually quite distinct, then L will not simply be formed by taking the union of the two disjoint vocabularies; L must at least contain additional concepts or rules which serve to link the given terms in a suitable manner; and some form of translation is almost certain to be involved. If, on the other hand, one starts with theories containing shared terms, one can assume that these occur either with the same meaning or not (the meaning of 'meaning' here is unimportant). In the first case, at least a partial translation will be available as an identity mapping on the shared expressions, or on that subset of expressions having the same meaning in either theory. In the second case, the existence of a common extension L that is at the same time semantically interpreted cannot generally be assured, due to the incompatibility of meaning displayed by some terms. At least it is quite unclear how such a language L could be in the semantical sense a *conservative* extension of the two languages one started with, i.e. one in which the original meanings are preserved.[1] In this situation — one which is characteristic of commensurability disputes — recourse to a direct intertheory translation seems to be indispensible, therefore, if rational comparability is to be achieved. And the assumption of translatability may, contrary to first appearances, be weaker than the assumption of a common, 'overarching', comprehensive language.

The idea that translation is the principal ingredient required for revealing rationality and objectivity in science is well entrenched in philosophical tradition. By the same token, the claim of nontranslatability, as applied to the expressions of different languages, frameworks and theories, has been a key and recurrent feature of various kinds of conceptual relativism over the years. To mention only some, it belongs to Le Roy's nominalism or extreme conventionalism, to Adjukiewicz's doctrine of radical conventionalism, to Whorff's theory of linguistic relativism, and of course to Kuhn's and Feyerabend's versions of the incommensurability thesis. Thus far, in isolating translatability as a necessary condition for having a reasonably robust type of comparability and commensurability, we seem to tread on fairly solid ground.

But I have also been arguing that under suitable conditions translatability can be seen as *sufficient* for commensurability. This idea requires further elaboration. Plainly, it would be absurd to claim that any arbitrary correlation of expressions counts as an adequate translation and serves up the requisite degree of cogni- tive contact between different scientific theories and conceptual frameworks. For this reason, one needs to determine what specific types of constraints on translation are in fact appropriate in this or that context.

As we have seen, this question can be approached in various ways. The general type of answer provided by standard semantic theory is that translation has to preserve appropriate semantic values, of which reference is perhaps the central one. Applying this constraint to relations between empirical theories, I have argued in Chapters 1 and 2 that translation can be considered adequate in the semantical sense if it respects an accepted model- theoretic reduction of the structuralist sort. Certain additional requirements for translatability, like Balzer's condition of literal- ness, were found to be inappropriate.

I also suggested in Chapter 4 that the relations of correspon- dence, interpretation and embedding, likewise provide reasonable standards of reducibility, hence of translatability and commensu- rability as well. However, there are controversial cases in the history of science where the applicability of such relations has been emphatically challenged. For these cases we may need to pay closer attention to the subtleties of translation, and to con- sider what special desiderata, if any, for adequate translations might apply. And, where possible, we should try to accomplish the task of translation.

I reviewed at the end of the last chapter some kinds of criteria for adequate translations one might want to apply in connection with the most celebrated example of this kind — namely, the comparison of classical mechanics and special relativity theory.

This example is an important one that can profitably be used as a test case. But still it embodies only some characteristic features of conceptual disparity in science. There are other features which equally demand our attention. Moreover, though I would argue that the methods for handling the CM/RM example can also be applied elsewhere in physics, they do not exhaust our resources for establishing comparability and commensurability. In the remainder of this chapter I shall also be looking, therefore, at some other approaches, and at the more general issue of how certain types of intertheory translation serve the aims of rational theory appraisal and choice, and how they may actually be available in certain cases. The outcome will be nothing like a blueprint for commensurability, nor a manifesto for unbridled rationality. It will amount to a brief assessment of the state of the art, together with some tentative proposals for signposting two general routes to commensurability.

One route to commensurability that I want to discuss is by way of a piecemeal 'local' analysis of primitive terms and basic sentences. The aim is to find common ground for intertheory comparison. The method is to dig deep enough into the conceptual soil of each theory so that eventually a layer of shared terms is unearthed. This process, of identifying common features, extrapolating and generalising from them, is a familiar and trusted technique of scientific method and theory construction. Historically, it is perhaps most closely associated with the conventionalist tradition in science and epistemology. These days, however, this approach is increasingly used in philosophical and foundational studies of physics, particularly in the areas of space—time theory and mechanics (classical, relativistic and quantum mechanics). Intertheory translation is an integral feature of the enterprise. Typically, the translations involved are *partial*, in that they do not act on all expressions of the translated language or theory; but the logical contact they achieve is, as a rule, especially solidly anchored. They come nearest to what could be termed 'theory-neutral' translations, though the neutrality is to be understood here in a relative and not absolute sense.

This approach comes in several variations, but there is one in particular I should like to draw special attention to: Haim Gaifman's theory of conceptual frameworks, as developed in his (1975), (1976) and (1984). Though it has an underlying pragmatic theory of meaning, Gaifman's approach is principally syntactical; matters of language and translation are in the foreground and questions of semantic interpretation are to some extent secondary. Moreover, Gaifman formulates quite specific adequacy conditions for translation, which, when satisfied, should establish tight conceptual and cognitive links between theories. And, unlike some ontological theories, this one has been applied (by Gaifman) to actual case studies, and the logical details rather fully worked out. As a consequence, they form a concrete basis for discussion of comparability and commensurability.

The second general route I want to discuss hunts commensurability at the surface. It makes a more global inspection of the terrain, and aims to set up a lexical and semantic correspondence between the languages of rival theories considered in their entirety. Unlike the first type of approach, here translation does not even have a limited theory neutrality. The point of reference is the target language itself as interpreted by a given theory, typically by a supplanting theory to which an earlier theory 'corresponds' or 'reduces'. The method is general, and is intended in particular to embrace those examples of scientific change where the later theory may be said to 'contain' its predecessor as a limiting case. The formal characterisation follows the contours outlined in earlier chapters under the rubric 'limiting case correspondence'. As an illustration of the method. I shall report some of the leading ideas of the case study of classical and special relativistic particle mechanics given in Pearce and Rantala (1984a; 1988). In particular, I want to review this analysis in the light of our discussion in the last chapter of the possible incommensurability of CM and RM resulting from the meaning variance associated with central terms like 'mass'. I shall try to show how the account which Rantala and I provide yields a new and viable way of clarifying the issue of comparability, and how it can be used to help resolve the conflict between the proponents of the doctrine of incommensurability and their critics. Briefly, the resolution consists in showing how to define a full but nonliteral translation from classical mechanics into the language of special relativity which satisfies those conditions of adequacy that we have already discussed.

2. ONTOLOGY AND CONCEPTUAL FRAMEWORKS

I have spoken often in these pages about the role of translation in forging the kinds of cognitive links between theories that are needed for their rational appraisal and, ultimately, for grasping the progressive character of science. Sometimes, due to changes of meaning, a nontrivial translation is needed even when the two theories under comparison share what seems to be the same vocabulary. In other cases, one has to compare and contrast two quite different systems of concepts or genera of explanations. The latter situation is common in the history of physics, where one finds, for example, caloric versus kinetic theories of heat, wave as opposed to corpuscular explanations for the transmission of light. And, in the more general context of quantum physics, there is the familiar 'duality' between wave and particle descriptions of atomic phenomena.

Such cases of apparently competing explanations based on different choices of concepts are often taken to typify differences of *ontology*. On this view, the photon and wave theories of light, for example, would simply constitute alternative, rival ontologies.

But the matter is not always that simple. Two *prima facie* conflicting theories might actually be, in a well-defined sense, equivalent; not only in their observational predictions, but even with respect to their ontic commitments. Mathematics possesses a plentiful stock of examples. A geometry based on points might, for instance, be equivalent to a geometry which takes line segments as primitive. Equivalence in this sense means that each theory is reducible to the other; more precisely, that they are mutually interpretable, or intertranslatable in such a way that each theorem of one system is transformable into, and in turn is obtainable from, a theorem of the other.

This idea of equivalent descriptions is not confined to mathematics, but extends to the physical sciences as well. Geometries may, of course, receive a physical interpretation; and space—time theories in general underlie large portions of both classical and relativistic mechanics. Besides the fact that geometry and space—time theories play a central role in physics, certain geometric *analogies* have been influential in discussions of the possible euqivalence of different theoretical concepts in science. The expression 'duality', for instance, that is widely used in connection with the wave-versus-particle descriptions of light or of matter, has a precise, formal counterpart in geometry and in axiomatic systems generally.

On the view of naive realism, then, conceptual differences reflect ontological differences and conceptual similarities are symptomatic of ontological similarities. From a more discerning, realist or nonrealist, stance, neither of these connections is watertight. Just as distinct systems of concepts can give rise to the same ontology, so ontological differences can sometimes be attributed to theories which are couched in languages that are, from the syntactic point of view, basically the same. This last point is not always easy to appreciate when one adopts a purely 'objectual' account of ontic commitment, of the Quinean sort that is popular these days. For, on that view, there can be no difference of ontology without a difference of 'entities' or 'objects' referred to. But there are alternative approaches to ontic commitment which bring out these differences more effectively. Before turning to one such approach, let me take up briefly some historical considerations.

In both science and the epistemology of science, there have been important, and historically related, traditions in which theory appraisal and theory change have been analysed by seeking, where possible, to resolve theoretical and conceptual conflict, and to bridge apparent ontological differences between frameworks. The instrumentalist tradition in ancient and medieval astronomy (the latter associated in particular with Osiander and Bellarmino) and the conventionalist tradition in geometry and mechanics (associated mainly with Duhem and Poincaré) spring immediately to mind.

Some philosophers (e.g. Popper) have, in effect, identified these two traditions, and regarded the conventionalism of Duhem's and Poincaré's kind to be merely a variety of (strong) instrumentalism. But recent studies in the philosophy of Duhem and Poincaré have shown conclusively that this view is mistaken, or that it is, at least, much too simplistic.[2] Moreover, it is often alleged that the *equivalence* that one might claim between different conceptual schemes in physics — e.g. the wave and particle theories of optics — can only be at best of a limited, 'observational' kind, and that *equivalence* in any stronger, theoretical sense can be upheld only under a strictly instrumentalist view of science. This argument is likewise fallacious. From the logical standpoint it fails to concede the possibility of intertranslatable theoretical languages. As a historical claim, it fails to do justice to the epistemological views actually held by, for instance, conventionalist philosophers who asserted in some cases the equivalence of rival theories.

Consider Poincaré's conventionalism. On the most plausible account available, defended and documented in detail by Jerzy

Giedymin (1982), Poincaré's epistemology of geometry originated from certain mathematical advances in geometry (in the early and middle part of the last century) due, in particular, to Gergonne, Plücker and Lie. Especially influential on Poincaré was the idea of *duality* within and between different systems of geometry (present already in the work of Gergonne and Plücker), and, more generally, the idea of equivalent, intertranslatable geometries that was brought out by Lie's theory of geometric transformations.

Poincaré himself further developed these notions and applied them in the study of relations between Euclidean and non-Euclidean geometries. Adopting the now-familiar group-theoretic approach, he was able to associate to certain rival (metric) geometries characteristic groups (of transformations), and was led by this route to his well-known assimilation of the choice of geometry to the choice of an appropriate group. An extension of the group-theoretic approach into the sphere of mechanics gave rise to Poincaré's (generalised) conventionalism in physics at large. The point that needs special emphasis in the present connection is this: when examining in general the question whether two rival, physical theories or principles are equivalent (so that a choice between them might be a matter of convention), Poincaré construed the cognitive content of a theory not solely as a function of its empirical consequences, but also in terms of its conceptual structure — in particular, in terms of the *form* of its equations; for example, their type, order and the group (of transformations) which they admit. For Poincaré, only observationally indistinguishable theories possessing structurally similar theoretical laws would count, therefore, as equivalent. (This feature has led Giedymin, for example, to redescribe Poincaré's instrumentalism as *structural realism*.[3])

Equivalence of this kind is illustrated in the case already mentioned of the wave versus corpuscular theories of light. Hamilton's method in geometrical optics can be used to cast both theories in the same form; not only by exhibiting a common, underlying 'observational' theory (characterised by the geometrical optics), but also by highlighting the formal similarities between the equations describing Maupertuis' principle of least action (belonging to Newton's emission theory of light), on the one hand, and Fermat's principle of least time (belonging to Huyghens' wave optics), on the other. Hamilton's method was, moreover, also generalised to dynamics, where it embodied the kind of 'geometrisation' programme for mathematical physics that was prominent in Poincaré's conception of the "physics of the principles" and in his generalised conventionalism.[4]

Do there exist rules or strategies for resolving apparent ontological differences? One method, favoured by conventionalist philosophers, is also a standard technique of theory construction in physics. One might perhaps characterise it under the slogan: *search for a more abstract or general standpoint*. Sometimes, different types of phenomena, and rival ontologies, can be *fused* through the discovery of a more general theory or conceptual framework which explains both or which contains each as a special case. On other occasions, it may be possible to achieve a partial resolution of ontological differences simply by retreating to a more basic or fundamental set of concepts. Thus, Euclidean and non-Euclidean metric geometries can be compared from the more general vantage point of projective geometry, and generalisations of quantum logic can be used to locate a common ground between classical logic and ordinary quantum logic. Different patterns emerge here. In some cases the generalisation process isolates a set of concepts that is *shared* by two different theories, besides revealing possible conceptual disparities. In other cases, one proceeds by giving a fresh reconstruction of the theories, using new primitives, hoping thereby to indicate a shared conceptual base for them. One works then from the 'ground level' upwards, so to say, building up the usual concepts on the basis of the shared, underlying terms.

These kinds of techniques inspire a good deal of contemporary foundational studies in physics, for example, on relational theories of space—time, and the comparison of different space—time theories. I mentioned already in Chapter 2 the work of Field and others in this area, concerned with the reconstruction of space—time theory on the basis of physically interpreted primitives. Thus, Manders (1982), for instance, treats point-objects and possible configurations of them, and employs translation techniques to show that physical theories postulating infinite totalities of abstract objects (e.g. space points) can be reformulated on the basis of a qualitative, Leibnizian, 'relational' view of space and time, whereby spatiotemporal concepts are construed in terms of relations between ordinary physical objects. Likewise, Mundy (1983) deals with physical relations between concrete objects (finite sets of point-particles). And, by defining embeddings into so-called *inner product* spaces, he demostrates how both Euclidean space and Minkowski space—time can receive a precise 'relational' reconstruction. Malhas (1984) defines a basic structure shared by classical and special relativistic space—time. It comprises possible particles and events, an incidence relation between them, and a distinguished collection of light particles (photons). Here, the leading ideas are derived from affine geometry, and, in particular, from certain duality principles which permit the interchange of points and lines (to yield the so-called *open-strip* representation of space). All these works contribute, in different ways, to the 'reduction' or the comparison of the usual ontologies associated with rival physical theories.

In the conventionalist tradition following Poincaré, there have also been more elaborate attempts to explicate ontological and conceptual differences between frameworks, focusing especially on questions of *meaning*. K. Ajdukiewicz, one of the leading figures in Polish analytical philosophy, developed in the early 1930s a systematic theory of language and meaning, on the basis of which he was led to a radically conventionalist epistemology containing a thesis of nontranslatability or (in modern terms) incommensurability.[5]

Adjukiewicz's conception of language involves syntax and *meaning rules*. The latter are pragmatic rules whereby certain (declarative) sentences of a language must be accepted by the users of that language, either unconditionally (*axiomatic rules*), on the grounds of other accepted sentences (*deductive rules*), or in the presence of certain perceptions (*empirical rules*). The meaning of a given expression is then specified through the (meaning) relations it has to other expressions of the language as laid down by the meaning rules. Ajdukiewicz also dealt with the 'open texture' of language, and with the idea of 'connectedness'. In his terminology, a language is *connected*, providing no subset of its expressions is unrelated (through meaning rules) to the rest of the language. And a language is *open* (or *not closed*) if new expressions, not equivalent in meaning to any existing expressions, can be introduced, without affecting the meaning relations already present in the language and without causing the language to become disconnected. A conceptual framework or 'apparatus' (*Begriffsapparatur*) was conceived by Ajdukiewicz as a closed, connected language in this sense.

One striking feature of Ajdukiewicz's approach to meaning is that translation between conceptual frameworks, or closed, connected languages, is an all-or-nothing affair. Two frameworks are either fully intertranslatable, or else completely 'disjoint'. This consequence of his construal of synonymy and translation seems to have been responsible for his claim that some frameworks in physics — e.g. the languages of classical and relativistic mechanics — are not intertranslatable, hence incommensurable. One might, however, revoke this thesis of nontranslatability, either on the grounds that real scientific languages are not genuinely closed and connected in Ajdukiewicz's sense,[6] or by questioning the assump-

tion that contact and communication between different frameworks can never be *partial*.

The view that conflicting ontologies can indeed overlap, or
contain common elements, and that translation can, accordingly,
be partial, belongs to Haim Gaifman's theory of conceptual
frameworks (Gaifman, 1975; 1976; 1984). Gaifman's approach
can, I believe, be located in or close to the conventionalist
tradition of Poincaré and Adjukiewicz; but, with regard to certain
aspects of language and meaning, it also draws on the philosophical tradition associated with Wittgenstein and Dummett. Like
Adjukiewicz, Gaifman adopts a pragmatic theory of meaning; like
Wittgenstein, he construes ontological commitment in terms of
the 'factually meaningful' or objectively true-or-false statements of
the language.

For Gaifman, a conceptual framework L comprises a language
(in the syntactical sense) together with sentential operators 'True'
and 'False' (where, e.g., 'True (φ)' simply means 'φ is true'), and
accompanied by a notion of *validity*, $\vdash_L$. It is assumed that the
users of L affirm certain 'evident conceptual expectations'; that is,
statements accepted as *a priori* truths, either unconditionally, or
on the basis of some deductive apparatus for proving sentences.
(These features thus resemble the axiomatic and deductive meaning rules of Ajdukiewicz.) Thus, if φ is a sentence accepted in this
way, one writes $\vdash_L \varphi$ and says that φ is L-valid. Likewise, users
of L regard certain statements as having factual meaning in the
sense that they are objectively true-or-false unconditionally. This
class of statements, possessing definite, though possibly unknown,
truth values, determines the ontological commitment of the
framework. For a sentence φ in this class, one may therefore
write: $\vdash_L \text{True}(\varphi) \vee \text{False}(\varphi)$.

Gaifman also defines open and closed frameworks (and further
grades of openness) in terms of the validity concept. However, the
distinction captured here is not that between having and not

having truth value gaps, since even for a closed framework some sentences may lack objective truth values. It is rather that, for every statement φ in a closed framework, the statement expressing the truth of φ can be associated with an equivalent statement which does have objective truth value.[7]

Translation from a framework L into a framework L' is conceived by Gaifman to be a recursive mapping t of L-sentences into L'-sentences. t need not be everywhere defined; that is to say, t may be a partial function. But the domain of t should include all sentences whose truth-or-falsity is L-valid, and the mapping itself should preserve validity and the ontic commitment of the frameworks, as well as respect the logical connectives (if the two frameworks share the same logic). For example, t should satisfy

$$\vdash_{L'} t(\varphi \wedge \psi) \leftrightarrow (t(\varphi) \wedge t(\psi));$$
$$\vdash_L \varphi \quad \text{iff} \quad \vdash_{L'} t(\varphi);$$
$$\vdash_L \text{True}(\varphi) \vee \text{False}(\varphi) \quad \text{iff} \quad \vdash_{L'} \text{True}(t(\varphi)) \vee \text{False}(t(\varphi));$$

and, furthermore, it is required that the language used for defining t have a minimal ontological commitment with respect to L and L'.[8]

Gaifman applies these ideas to various examples of closed frameworks. For instance, he represents Newton's (absolute) space—time ontology by means of several different but equivalent frameworks, for each of which there is a precisely defined counterpart framework representing Leibniz's (relational) space— time ontology. In every case, there exists a translation in the above sense from the Leibnizian framework into the appropriate Newtonian one. And by a suitable choice of framework, this translation can be chosen as an identity mapping (on *some* sentences); translation is therefore partial, but, to that extent, literal. In this manner, Leibniz's viewpoint is reconstructed as an affirmation of a certain "simply defined sublanguage of the

Newtonian framework" (Gaifman, 1976, p. 65). Irreconcilable ontological differences remain, however, because no full translation is available which would render the two frameworks equivalent. The partial translations establish a genuine divergence of ontology, but at the same time they pinpoint an area of common ground sufficient for counting the frameworks as commensurable.

Though Gaifman's approach is predominantly language oriented, it does not preclude the possibility of incorporating a structural, semantic component. In fact, Gaifman (1984) includes a model-theoretic description of the Leibniz—Newton example. In this case, the characterisation given bears more than a passing resemblance to the general standpoint adopted by Poincaré which I mentioned earlier. For, the classes of statements representing respectively Newton's ontology and Leibniz's ontology, though formulated in the same language (syntax), are demarcated precisely by means of (transformation) *groups* under which they are invariant. In terms of structures, the sentences of each class preserve their truth values under the appropriate mappings (transformations) of models. Thus, at least for this example, one can infer that Gaifman, like Poincaré, assimilates the choice of ontology to the choice of a particular group.

3. THE TRANSLATION OF CM INTO RM

I turn now to the second route to commensurability, which will be illustrated by sketching, as promised, the way in which the language of classical particle mechanics can be translated into the conceptual framework of special relativity. I should emphasis that what follows *is* merely an outline of the translation and its chief properties; and I shall gloss over many of the technical details involved. The reader interested in the full story should consult Pearce and Rantala (1984a) and, especially, (1988).

To start with, let us assume that CM and RM represent the core theories of classical and relativistic mechanics, based respectively on Newton's second law and the Minkowski force law. Accordingly, let s denote the spatial position of particles, let m stand for their mass and f for force. These, together with the requisite mathematical terms, will be the shared primitives of CM and RM; the latter theory has additionally c as an individual constant denoting the speed of light. Thus, in simplified form CM is characterised (essentially) by Newton's law:

(1) $m \cdot d^2s = f;$

and RM has as its central law:

(2) $m \cdot d(ds/(1-(ds/c)^2)^{1/2} = (1-(ds/c)^2)^{1/2} f;$

where 'd' denotes the derivative with respect to time. Clearly, in the relativistic equation m plays the role of rest or proper mass.

The correspondence of CM to RM can, of course, be described mathematically by the property that (1) is obtainable from (2) in the limit $c \rightarrow \infty$ or in the limit $ds/c \rightarrow 0$. However, the problem of formulating a semantically adequate and precise *logical* relation between the two laws is exacerbated, as we have already seen, by the following three factors:

(i)   In view of the alleged conceptual disparity, it seems as if 'm', and possibly other terms, have different meanings in (1) than in (2).

(ii)   If one neglects the case of zero particle velocities as trivial and physically insignificant, then (1) and (2) are *prima facie* incompatible, making the explanation or deduction of one law from the other problematic.

(iii) The limiting assumptions required if one is to derive (1) from (2) seem to be, in the first case, counterfactual (or inconsistent with the empirical claim that c is a fixed, finite number), in the second case, satisfied only in the uninteresting domain of stationary particles.

The central task, then, is to establish an appropriate logical relation between CM and RM which preserves the basic intuition behind the claim that (1) and (2) 'correspond in the limit' (in so far as this intuition *can* be coherently preserved), and, at the same time, takes account of the three difficulties just indicated. In order to motivate the solution to this problem, and to ground the translation that arises from it, let us consider the situation model-theoretically.

When one thinks in terms of semantic structures, the first thing one notices about (1) and (2) is that certain models of RM are very close to certain models of CM. 'Very close' here means roughly this: the values (real numbers, vectors, etc.) of position, mass and force assigned to each particle in the one model are numerically close to the values assigned in the other model. For example, a relativistic model which represents a system of 'slow-moving' particles over a given period of time will assign values for s, m and f which are numerically close to those values assigned by a structure representing the same system of particles but satisfying equation (1) instead of (2); and the situation is completely analogous if, in the model of RM, c has a 'very high' value. One might put the point a little more colourfully by saying that some relativistic 'worlds' are *almost* classical.

This fact is, of course, very well known. What is less obvious is how one should exploit it to define a semantic relation between the theories by correlating those models of RM with their proximate counterparts in CM. The solution to this problem was first formulated in a paper by Veikko Rantala (1979). Rantala's idea was to use nonstandard analysis, or the theory of infinitesimals, to explicate in a logically precise manner the sense in which certain models of RM are 'close' to counterpart models of CM, in fact *infinitesimally close*. Subsequently, Rantala and I extended this method to show in effect how the subjunctive, counterfactual conditional, under which this portion of classical mechanics can be said to be entailed by relativistic mechanics, can be replaced by an equivalent, indicative, truth-functional condition that is compatible with RM. If β denotes the ratio ds/c, and the symbol '$\approx$' stands for 'is infinitesimally close to', then the condition in question can be informally written as '$\beta \approx 0$'. The logical relation of RM to CM can then be expressed by saying that the conjunction of (2) and ($\beta \approx 0$) entails a *translation* of (1); hence it takes the form of a deductive explanation of CM from RM. Naturally, all this has to be made precise by formalising the equations and limiting condition involved, and by defining a suitable translation between the formal languages. This, very briefly, proceeds as follows.

Let the vocabulary of CM be specified by a type τ, where, in particular, τ contains the symbols s, m and f as well as names for all the mathematical objects and relations one needs. Suppose that τ is three-sorted and that variables x, p and t are used to range respectively over mathematical objects (e.g. reals), particles, and time points. One can take the type of RM to be $\tau' = \tau \cup \{c\}$. Let L be a sufficiently strong (say, infinitary) logic, so chosen that in $L(\tau)$ one can axiomatise by some sentence θ the class M of models of CM.[9] In particular, structures in M must satisfy (1) together with various mathematical assumptions of continuity, differentiability, etc., and they should consist of finite sets of particles. Likewise, assume that the class M' of models of RM (satisfying (2)) is characterised by a sentence θ' of $L(\tau')$. Thus, elements of M can be taken to have the form:

$$\mathfrak{M} = \langle \mathfrak{A}; P, T, s, m, f \rangle$$

where $\mathfrak{A}$ is a purely mathematical, standard or nonstandard model of analysis, P and T are domains corresponding to sets of particles and time points, and s, m and f are the appropriate functions. Models in M' will be of the form $\langle \mathfrak{M}; c \rangle$.

The next step is to define by a suitable $L(\tau')$-sentence ψ, say, a subclass K' of M' comprising relativistic models

$$\mathfrak{M} = \langle *\mathfrak{A}; P, T, s, m, f, c \rangle$$

such that $*\mathfrak{A}$ is a nonstandard model of analysis, velocities and accelerations in $\mathfrak{M}$ are very small compared to c (that is: the ratio is infinitesimal), and T, s, m and f satisfy certain finiteness properties which make them 'nearly' standard (more precisely they possess *standard* parts).[10] Then, given any model $\mathfrak{M}$ in K', let $F(\mathfrak{M})$ be the structure which results from $\mathfrak{M}$ by (i) replacing $*\mathfrak{A}$ by a standard model of analysis $\mathfrak{A}$, where $*\mathfrak{A}$ is an elementary extension of $\mathfrak{A}$; (ii) replacing T, S, m and f by their standard parts; and (iii) deleting c. F can be defined as an operation or construction which yields, for each $\mathfrak{M}$ in its domain K', a unique model $F(\mathfrak{M})$ that one may call the *standard approximation* of $\mathfrak{M}$. Evidently, $F(\mathfrak{M})$ is a structure of type τ. It can also be shown that $F(\mathfrak{M})$ belongs to M. In fact the range of F consists roughly of all 'standard' models of CM in which velocities and accelerations are bounded by finite real numbers, and accelerations are continuous. If L is appropriately chosen, these properties will also expressible in $L(\tau)$, so that the range of F can be defined as a subclass K of M axiomatised by a suitable $L(\tau)$-sentence ζ.

The function F thus provides a formal characterisation of the feature noted earlier, that certain classical and relativistic models are 'very close'. It determines, therefore, the required semantic correlation of the two theories by uniquely associating models of CM to given models of RM. The final step is to define a translation which respects this correlation in the proper manner.

To begin with, consider two fixed L-formulas, $\gamma(t)$ and $\sigma(x)$. The first should formalise the condition that t is a point belonging to a standard time interval; the second that x is a standard mathematical object (real, function of reals, etc.). Now, let φ be any atomic formula of $L(\tau)$ of the form $u_1 = u_2$, where u_1 and u_2

are terms of any sort. Define $I(\varphi)$ to be the L-formula which says that $u_1 \approx u_2,$[11] if u_1 and u_2 are nonstandard numbers, or which sets $u_1 = u_2$ if they are both particles or standard numbers. Define $I(\varphi)$ analogously if φ is of the form $u_1 \in u_2$. Extend I inductively by setting:

$$I(\neg\varphi) = \neg I(\varphi)$$
$$I(\wedge\Phi) = \wedge\{I(\varphi): \varphi \in \Phi\}$$
$$I(\forall t\varphi) = \forall t(\gamma(t) \rightarrow I(\varphi))$$
$$I(\forall x\varphi) = \forall x(\sigma(x) \rightarrow I(\varphi))$$
$$I(\forall p\varphi) = \forall p I(\varphi).$$

I is thus a recursive and nonliteral translation of all $L(\tau)$-formulas, with quantifiers relativised at the appropriate places. It is a fairly straightforward matter to show that I is an adequate translation in the sense that for all $\mathfrak{M} \in K'$, for all $\varphi \in \text{Sent}_L(\tau)$.

(3) $\mathfrak{M} \vDash_L I(\varphi) \Leftrightarrow F(\mathfrak{M}) \vDash_L \varphi.$

The translation does, therefore, respect the semantic correspondence. To give a simple example, suppose that φ is an atomic expression involving 'mass', e.g. a sentence of the form $m(p) = r$ which states that the mass of a particle p is equal to the real number r. The relativistic translation of φ is a sentence $I(\varphi)$ (not necessarily atomic) which states roughly that the (rest) mass of p is infinitesimally close to r. Condition (3) then ensures that the translation of φ is true in a model in K' just in case φ itself holds in the standard approximation of that model. Moreover, since we have defined K' to be an axiomatisable subclass of M', it follows that any sentence φ which is true classically (and thus holds in particular in all models in the range K of F) has a translation $I(\varphi)$ which holds in all models in K', and is therefore a semantic consequence of the conjunction of θ' with the 'limit condition' ψ. In other words, if φ is a classically correct statement, its translation is relativistically valid in the limit of low velocities. More to the point, Newton's second law (1), suitably formalised, is, under translation, a logical consequence of RM under the auxiliary assumption $\beta \approx 0$. In general, the relation between the two theories is expressed by

(4) $\theta', \psi \vDash_L I(\theta)$;

showing that in the limit CM can be deductively inferred from RM under this translation.

The satisfaction of (3) also determines a sense in which the explanatory power or problem-solving ability of CM is transferred to RM, so that relativistic mechanics can be said to retain the successful part of classical mechanics. The argument for this follows the pattern of Chapter 5. If one interprets the class K' as representing the 'domain' of validity of CM as seen from within RM, then from the relativistic perspective an explanation in CM of a statement φ will have the form:

$$\zeta, \xi \vDash_L \varphi,$$

where ξ formalises the initial conditions and ζ axiomatises the F-image K of K', as before. Since I is truth-preserving relative to F, one can infer that

(5) $\theta', \psi, I(\xi) \vDash_L I(\varphi)$.

In other words, in the limit $\beta \approx 0$, represented here by ψ, RM explains the translation of φ under the same initial conditions (translated into the language of RM). With the rider that K' be construed as determining the limits of validity of CM, it seems reasonable therefore to claim that RM *retains* the successful part of CM. Naturally, this conclusion holds good for these core fragments of classical and relativistic mechanics; the theories that might, in the terminology of the Chapter 5, be taken as bases for reconstructing the classical and relativistic research traditions as theory ensembles. It does not, of course, imply that the explanatory power of *all* branches of classical mechanics is matched within the special theory of relativity. To study this problem, one would have to give a precise characterisation of both ensembles and to extend the correspondence relation accordingly.

4. EXPLANATION AND MEANING

The foregoing analysis provides an indication of how CM and RM can be rationally compared. It shows that these two theories are, after all, conceptually and cognitively related. They are, in short, *commensurable*.

The problem of establishing a suitable connection between conceptually disparate terms, like 'mass' in CM and RM, is thus resolved by the expedient of translating all occurrences of the classical terms into relativistic language. At first sight, this translation might seem artificial and contrived. It is, admittedly, available in virtue of a rather special formalisation of the languages involved. And it is semantically adequate, but with respect to a correlation of some, and not all, models of the two theories. These properties might initially cast a shadow over the success of the rational reconstruction. But any doubts that might arise can. I think, be dispelled by showing that the translation is in fact compelling, and does provide an adequate response to the challenge of Kuhn and Feyerabend.

In the first place, it is important to notice that the translation is indeed a recursive mapping of all expressions, and is definable within the apparatus available to us in the source language. It is therefore not open to the kind of objection I raised earlier against Levin's proposal. This does not mean, incidentally, that each occurrence of 'mass', say, is translated in exactly the same way. How 'mass' is to be translated depends on the make-up of the sentence which contains it. But it is determined in a rule-governed way by the syntactic structure of the given sentence; it does not appeal to extrinsic properties like a truth predicate for RM, nor is it settled in advance of the translation process what the truth value of the outcome is to be.

What the translation preserves are suitable semantic values, given the association of classical with relativistic structures. The fact that not *all* models of the two theories are correlated should not prejudice the propriety of the ensuing translation. It is merely a reflection of the feature that not all classical and relativistic 'worlds' *are* sufficiently 'close' to warrant the association. One's attention is thus directed precisely at the 'limiting' models of RM and their relation to their classical counterparts. Once the correct connection is established at the semantical level, the form of the syntactic translation is virtually settled. Its acceptability, moreover, cannot be called into question on the grounds of its being a highly *formalised* mapping of expressions. On the contrary, this property may help to explain why agreement on the rational comparability of CM and RM has been so hard to achieve at the informal level of analysis, and why a more sophisticated logical reconstruction may achieve better results.

How does the translation fare with respect to the difficulties I summarised in the last section, namely the problems of 'meaning variance', 'incompatibility' and 'counterfactuals'? The standard accounts seem to run aground when trying to handle simultaneously these three features of the CM/RM relation. In fact, it is quite common practice to ignore one or more of them, or else concede that all three together cannot be properly accounted for in a rational reconstruction of the transition from classical to relativistic mechanics. Either way, this is grist to the incommensurablist's mill. It either allows him the rejoinder that an important aspect of this intertheory relation has been neglected, or else it is liable to reinforce his own thesis, as when, for example, translatability is the key property that the rationalitist gives up.

The analysis I have been sketching shows quite plainly, however, that all these features belonging to the CM/RM relation can be adequately dealt with. The resolution is even rather simple; once one feature has been correctly diagnosed and dispatched, the others slot neatly into place. One observes, at the start, that CM and RM *are* to all intents and purposes syntactically incompatible, besides being conceptually disparate. By translating the language of CM *non*-literally, the acknowledged differences in meaning are *corrected for*, and one is able to define a precise logical relation between RM and CM *re-expressed in relativistic language*. This logical relation is no longer one of simple incompatibility (which held between the semantically unanalysed theories), but rather of *derivability in the limit*. Lastly, this method shows that the character of this limit condition — whether it is consistent with RM, or counterfactual and incompatible with RM — depends crucially on the way in which the two theories are reconstructed. By using the theory of infinitesimals, one is able to provide a quite natural reconstruction under which the classical limit of RM is actually compatible with RM (even for nonstationary particles). What is more, this is no *ad hoc* device. For, one may justifiably claim that the new approach to mathematical analysis which Abraham Robinson's theory opens up is a rigorous reformulation of the basic ideas already developed by the founders of the calculus, ideas that have essentially been retained ever since in intuitive thinking about limits and approximations.

This last feature of the reconstruction, namely its treatment of allegedly counterfactual conditions, may have a pertinent role to play in responding to Feyerabend's arguments for incommensurability. He gives surprisingly few *concrete* grounds for rejecting translation in cases like CM and RM. More often that not his critique is aimed at underscoring the phenomenon of meaning variance by pinpointing differences in properties that classical and relativistic concepts possess. Since these differences do not in principle constitute a barrier to translation, the question arises: why *does* Feyerabend insist so emphatically on nontranslatability? I have no simple answer to hand. But perhaps some clues may be gathered from a passage which follows one of his characteristic statements of the conceptual differences between classical and relativistic terms.

> It is also impossible to define the exact classical concepts in relativistic terms or relate them with the help of an empirical generalization. Any such procedure would imply the false assertion that the velocity of light is infinitely large.   (Feyerabend, 1962, pp. 80—81)

This line of argument is not easy to evaluate because Feyerabend does not indicate how a term like Newtonian 'mass' *would* be relativistically definable if the velocity of light were infinite, and undefinable otherwise. But what does emerge quite clearly is this. If Feyerabend's main concern is that the classical limit of RM may require a counterfactual characterisation, then our reconstruction seems to offer an adequate rebuttal. For it shows how classical *sentences* can be translated without assuming that c is infinite, or at least without forcing that assumption to be incompatible with RM. And it is the relativistic consistency of this condition or the more general $\beta \approx 0$, rather than its truth value in some absolute sense, which appears to be decisive for obtaining a reasonable logical relation between the two theories.

As we have seen, this relation can be expressed in the form (4) of a deductive explanation of CM by RM.[12] Moreover, in this form it is no longer susceptible to objections of the sort which Feyerabend (1962) raises against the standard concept of inter-theory explanation (of the Hempel—Nagel kind) or the slightly revised version of explanation that he attributes to Popper and Watkins. According to the latter account, a theory T' may explain its predecessor T even though they are mutually inconsistent. It is assumed that there is some domain D, say, in which the two theories are *empirically* indistinguishable, though formally incompatible, and that T' corrects T whilst showing it to be approximately adequate within D. Thus T' is said to explain T under the boundary conditions which define D, and, of course, in this sense it retains the valid part of T.

Feyerabend's attack on this position strikes me as basically sound. He remarks that if T and T' are incommensurable, their classes of observational consequences will be disjoint and noncomparable, hence the theories will be unrelatable for content. If, on the other hand, T is simply inconsistent with what T' asserts within D, then the resulting notion of explanation is bound to be inextricably tied to available observational methods, since these alone determine what is and what is not experimentally distinguishable. Both situations are unsatisfactory for achieving *bona fide* explanations, but it is readily seen that neither case fits the pattern of our schema (4). In the first place, according to (4), CM and RM are rendered comparable by translation. And, secondly, under the condition $\beta \approx 0$ which defines the appropriate domain D, RM and (the translation of) CM are *exactly* and not merely experimentally indistinguishable.

This last property indemnifies our characterisation from the kind of criticism Feyerabend levels against the Popperian view of intertheory explanation and it justifies our speaking of a genuine relativistic explanation, in the deductive sense, of Newton's law. Naturally, that law remains relativistically valid only *in the limit*, but one need not interpret this limit as being inconsistent with RM. Analogously, schema (5) indicates that a class of Newtonian problem solutions is also relativistically valid in the limit. This transfer of problem-solving ability from CM to RM should be understood with some caution, however. In general, (5) is not to be read as claiming that RM reproduces the *same* solutions to given problems. In fact it is well known that as a rule RM alters the classical interpretation of an empirical problem and offers up a new solution of its own making. The point about translatability is rather that classical solutions have a *representation in the relativistic framework*. This representation permits the *comparison* of classical and relativistic solutions; and (5) then shows that certain classical answers to problems are, by the lights of RM, valid under the boundary condition $\beta \approx 0$. It also shows, of course, that RM can be used to evaluate the adequacy of classical solutions and establish the appropriate restrictions on their validity. In this sense RM functions so as rationally to *appraise* CM.

The proposed reconstruction thus indicates how one can preserve the central requirement of the rationalist account of scientific progress, namely that the later theory may serve as a vehicle for the cognitive appraisal and reappraisal of its predecessor. Yet this is possible in spite of the fact that the supplanting theory may introduce a fundamentally new conceptual framework, that some or all terms of the old theory may undergo a shift in meaning, and that experimental findings may receive a new interpretation accordingly. An adequate theory of rational progress need not, and indeed should not, be wedded to the ideal of stability of meaning of reference; nor, conversely, does it follow from the 'incompatibility' and 'meaning variance' hypotheses that scientific change cannot be grounded according to the usual canons of rationality.

This, at any rate, is the view which I think can be maintained in the case of CM and RM, and almost certainly for other examples of conceptual change where the theories involved correspond in the limit. Clearly, this claim depends on the availability of a suitable translation. And since this is such a crucial point, it is also the one most likely to come in for criticism. Let me therefore try to anticipate some of the sorts of objections that might be brought to bear on it.

Perhaps the principal question to consider is whether the translation of CM into RM requires any additional support or justification over and above those features and properties of it that I have already described in outline. In general, the demand for justification is a natural one. But in this case I think one can reply to the challenge with another question: What kind of justification *would* be convincing here? It seems fairly obvious that *empirical* evidence is not the kind of support one is looking for. One is not, after all, setting up any explicit bridge principles linking the conceptual system of classical mechanics with that of relativity theory, principles of the sort that one might seek to ground on the basis of independent experimental results. Perhaps, then, one needs further *semantic* support to 'verify' that the translation really preserves meanings?

In reply to this, one should observe first that the translation does not, and need not, appeal to any kind of theory independent *reference*. Unlike the stability of reference theorist, our goal is not to establish for the classical terms some sort of external, 'objective' reference, to which one then tries to attach relativistic names. Far from being a weakness of this approach, it seems to me to be a positive advantage that it deals basically with the theories themselves and the claims they make about their respective domains. On the other hand, our translation can be said to 'preserve reference' given a certain association of classical and relativistic *models*. For those who like to think in terms of semantic structures, or even possible worlds, and who find this particular association compelling, the adequacy of the translation is already settled. Others may prefer to take the translation and the structural correlation *together* and evaluate them jointly as a package.

In the present context it may be that the appeal to 'meaning', even theory-dependent 'meaning', is to some extent misleading and counterproductive. On one level it is fine to say that a good translation should preserve meaning; and sometimes it may even be possible to verify that it does so. But in the context of translations between scientific theories, excessive appeal to 'meaning' may give rise to a false impression that the translator has at his command a set of independently discovered 'meanings' which he somehow matches together piece by piece until the correct translation falls into place. Scientific theories are generally not equipped with the kind of precise meaning theory that would allow this procedure to work. The fact that there may be good reasons to think that a term like 'mass' changes meaning in the transition from CM to RM does not imply that one can point out some unequivocal 'meaning' for 'mass', that one can define its meaning in some absolute sense.

It seems much more appropriate and fruitful, therefore, to think that the processes of prescribing meaning and providing a translation go hand in hand together. One determines the meaning of Newtonian 'mass' in so far as one is able translate (occurrences of) it into relativistic language. In this sense, meaning determination is not *prior* to translation, but another side of the same coin. If pushed to provide some 'ultimate' grounding for the translation process, I think one need only appeal in this case to the correspondence relation itself, to the way that physicists actually came to understand classical mechanics in the light of the relativistic 'revolution'. As the relativistic understanding of space, time and mechanics arose one was able to appreciate the scope and limitations of classical theory, to assign a suitable niche for CM and to retain it as a limiting case. CM was simultaneously corrected and explained from the vantage point of RM.

In this way, one may turn around the demand for an independent justification of translation based on meaning, and argue that the translation and explanation of CM *establishes* the classical meanings in so far as they can be established at all. Once again, it would be senseless to ask for priorities here, to inquire whether the explanation determines the translation, or the translation the explanation. Much better to say that the appropriate interpretation of CM as relativistically valid in the limit leads simultaneously to translation *cum* explanation. Expressed in these terms, I think it is clear that the translation requires no additional foundation above and beyond those central properties of it, formal and methodological, already described. The translation can then be seen as the natural outcome of a three-stage process. The first stage is characterised by the recognition of a physically meaningful and intuitively acceptable *correspondence relation* between classical and relativistic mechanics. At the second stage, there is an informal, syntactical representation of this relation which is open to the charge of being 'purely mathematical' and of failing to reproduce such features as 'meaning variance' and the deductive explanation of one theory by the other. The third stage consists in turning the informal correspondence relation into a formally precise *conceptual* link which gels with the physical interpretation of the relation and at the same time provides a coherent account of meaning change and explanation.

In the last resort, a logical reconstruction of theory change has to be judged by the manner in which it meets certain standards of accuracy and by the pragmatic norm of overall fruitfulness. As regards the pragmatic issue, one should, I believe, favour those reconstructions which reveal as far as possible rationality and continuity in the development of science, without obscuring the nature of the fundamental conceptual changes which are characteristic of scientific growth.

5. SCIENTIFIC CHANGE AND RATIONALITY: SOME TENTATIVE CONCLUSIONS

The two routes to commensurability reviewed in this chapter have to some extent contrasting aims and properties. Gaifman's approach, and variations on its general theme discussed in Section 2, is directed chiefly at locating ontological similarities and differences, and it leads principally to intertheory translations that are partial but literal. The approach to correspondence discussed in the last two sections embodies full but nonliteral translatability. It is evident, however, that no real conflict arises between the two methods, because one is likely to succeed in circumstances where the other would be inapplicable or would reap poor results. The first approach is the more natural one to pursue when there are good chances for generalising the ground-level terms of two theories so that a common set of underlying concepts and assumptions can be isolated. The second is better suited to those cases where two theories share a major portion of vocabulary, but where one of them has extra parameters whose limiting values turn that theory into a close approximation of its rival. Where comprehensive physical theories are concerned, it is likely that both methods can be applied in order to build up an accurate, global picture of intertheory relations.

I mentioned earlier that the second approach, unlike the first, can lay few claims to being 'theory neutral'. This term is a vague one, but I think one can now see the difference as consisting in roughly the following. Gaifman envisages the translator as someone who 'stands between' two conceptual frameworks without fully committing himself to either. Though he is not, and perhaps even cannot be, without ontological and epistemic commitments, with respect to the comparison in question his stand is a neutral one. By constrast, the translator who seeks a complete re-expression of CM in the language of RM takes the relativistic perspective as his own. His understanding of CM, and his consequent translation of it, is highly indebted to his relativistic understanding of physical phenomena.

Cognitive comparisons of the 'neutral' sort are useful and informative when they are available. I doubt that they are always available; indeed, I argued in the last chapter that the relativistic perspective was the natural and perhaps the only stand to take when trying to grasp the conceptual change occurring in the case of terms like 'mass'. It seems to me that the demand by a number of philosophers of science for a 'neutral' stance on such questions has obscured the main issues. Both the incommensurablists and their critics have, on occasions, made the facile identification of rationality and objectivity with 'theory neutrality'. Kuhn has at times made the fallacious inference from 'nonneutrality' to incommensurability. And even a 'liberal' critic like Laudan, sworn to the theory-dependence of empirical problems, has sought to resolve the issues of rationality and irrationality by the subsequent ploy of granting empirical problems an autonomous life of their own.

Theory neutrality, then, is an admirable objective, but a rational account of scientific progress that is irreversibly attached to it can scarcely deliver the goods. But, I contend, the neutral standpoint is not the only one by which rival theories *can* rationally be appraised. In the case of CM and RM treated here, the essential point is that classical laws and problem solutions can be meaningfully expressed and evaluated in the relativistic framework. I think one can agree, therefore, with Kuhn and Feyerabend that a neutral comparison of the theories' contents does not exist. But, by denying their claim of nontranslatability, one need not infer that therefore no ordinary rational comparison — of PSE, explanatory success, or whatever — is available. The comparison proceeds from the relativistic vantage point; but this is exactly what one would expect. The fact that translation is a one-way process, that the classical physicist cannot fully evaluate relativistic claims from his own (classical) perspective, seems perfectly reasonable. He must indeed, to use Kuhn's expression, *learn a new language*, before he can assess the real merits of the new theory. But having learnt the language, he can, through translation (or an equivalent but perhaps less explicit cognitive process), come to a rational evaluation of both theories.

Does this mean that the prior acceptance of RM is a precondition for the translation and rational criticism of CM? I think not. It seems to me that the criticism of CM and the development of the new framework proceed alongside each other. The scientist may therefore *entertain* RM and employ its conceptual system to interpret and evaluate classical theory, without necessarily being *committed* to relativity theory in the epistemic sense. After RM has been *asserted* by the scientist, it can then be used to assign truth values (or degrees of belief, or whatever) to sentences of CM, now understood of course in the translated sense. This makes it possible to interpret scientific *choice* as a thoroughly rational process, even if it need not be in any obvious sense a rule-governed procedure, and even if not all steps in the process are fully and explicitly reconstructed at the time.

The main conclusions relevant to the rational view of scientific progress, as they emerge from the last two chapters, can briefly be summarised as follows. In the case of classical and relativistic mechanics, and analogous instances of intertheory correspondence, the presence of a translation permits a straightforward 'solution' to the 'paradox of meaning variance'. This solution allows one to accept, at least for the sake of the argument, the major premises on which the paradox is constructed: namely, that the meanings of scientific terms are determined by the whole theoretical context in which they occur, and that therefore terms occurring in two *prima facie* incompatible theories may have different meanings. One can even grant the further assumption that sentences of different theories involving such terms cannot directly stand in any semantic relation of entailment, incompatibility, etc.[13] However, one denies the main conclusion that observational evidence cannot decide between such theories and that their empirical contents cannot be logically compared. The correct comparison proceeds by *translating* sentences of the one theory into the language of the other.

Thus, central terms, like 'mass', belonging to classical mechanics have different meanings in the special theory of relativity. Experimental evidence which conflicts with the laws and assumptions of CM may be reinterpreted in the framework of RM so as to confirm RM. However, sentences of CM can be translated into the language of RM so that the classical laws are expressible in the new language. In fact, under this translation the laws of CM are seen to be relativistically derivable in the limit. In this sense RM explains CM and retains its valid part as a limiting case.

It would be excessive to maintain, on the basis of the evidence so far gathered, that *all* scientific change is unfailingly well founded and rational. Even if one takes the more reasonable and moderate view that science is in the long term progressive, it seems to me that rationality needs to be demonstrated for individual cases and episodes. And this, I believe, requires the combined effort of historical analysis and logical reconstruction to make explicit the cognitive intertheory relations obtaining. On the other hand, one may plausibly argue that examples like the transition from classical to relativistic mechanics are representative of a certain *pattern* of theory or paradigm change: the pattern of theory change accompanied by *correspondence* and retention of the earlier theory *in the limit*. Likewise, the kinds of examples mentioned in Section 2 above may well be typical of other types of theory development patterns: the fusion of different ontologies into a more fundamental one, and the resolution of ontological differences by a process of generalisation.

It remains to be seen how widely the methods I have described here are applicable, and whether the historically most significant episodes of scientific change conform by and large to these patterns. However, the fact that at least some of the most intransigent and philosophically controversial cases are amenable to this sort of rational analysis strikes an optimistic chord.

NOTES

NOTES TO CHAPTER 1

1. For a critical discussion of the logical foundations of the structuralist conception, see Rantala (1978) and (1980). A general assessment of Stegmüller's nonstatement view is to be found in Pearce (1981). Chapter 6 of Niiniluoto (1984) (based on a paper written in 1978) gives a thorough survey and analysis of the structuralist approach, especially as applied to the development of scientific theories and to Kuhn's philosophy; the bibliography of this work includes a quite comprehensive list of primary and secondary literature devoted to the structuralist view.

2. As compared with his informative discussion of Kuhn's writings, Siegel's summary dismissal of Stegmüller's reconstruction and 'defence' of Kuhn is disappointing. He fails, for example, to distinguish between (i) philosophical positions actually held by Kuhn, (ii) views attributed to Kuhn by critics such as Scheffler and Popper, and (iii) Kuhn's own assessment of what his views entail. He also ignores that fact that Stegmüller is in places *critical* of Kuhn, and thus attempts to correct what he sees as mistakes in Kuhn's account of science; whilst at the same time Stegmüller is wary of much of the critics' reaction to Kuhn, and tries therefore to reproduce the defensible core of Kuhn's theory in a sympathetic light so that at least part of the critics' objections will no longer be telling.

Thus, for instance, when Stegmüller claims, according to Siegel, that "normal science is not irrational" (Siegel, 1980, p. 372nn), he is referring to a central property of the structuralist conception of theories, namely, that as mathematical structures theories are immune to falsification because they are not the kinds of entities which *could* be said to be falsifiable. Here Stegmüller is applying his new metascientific paradigm to support Kuhn's conclusion as to what his (Kuhn's) theory entails against the critics' analysis of what his (Kuhn's) theory entails. On the other hand, when Stegmüller asserts (also according to Siegel, *ibid.*) that "the criteria of evaluation of competing paradigms are not paradigm-bound", he is explicitly endorsing what some of Kuhn's critics (e.g. Scheffler, to whom Stegmüller refers) have pointed out: that there may exist *external* or *metatheoretic* criteria on the basis of which competing paradigms could be compared and evaluated. Siegel merely lumps together the different components (supportive or corrective) of Stegmüller's reconstruction of Kuhn and concludes that it concedes the critics' main arguments, and thus "fails to clarify or improve Kuhn's position".

I cannot say whether Siegel's misrepresentation of Stegmüller is typical, but it is possible that it does reflect a tendency on the part of some philosophers of science to be sceptical over the introduction of new formal tools and frameworks in metascience and hence to be overzealous in condemning them.

3. In the structuralist literature reduction is usually represented by a relation on $M \times M'$, whose converse is a partial function from M' to M. Here and below, I work directly with this converse function ρ.

4. The foregoing sketch is not intended to do justice to the full complexity of the structuralist view, but to bring out only some basic features required for the analysis to follow. For authoritative treatments the reader should consult Balzer and Sneed (1977/78). Stegmüller (1973/76; 1979, especially the Appendix, and 1986).

5. The first of the above three objections has been raised, for isntance, by Feyerabend (1977) and Pearce (1981). Sneed's reduction concept has been criticised and amended from a structuralist standpoint by Mayr (1976). And various aspects of the Kuhn-reconstruction are challenged by Niiniluoto (1984).

6. Schaffner (1967, p. 141) correctly reports Suppes' criterion of reduction and supplies a standard definition of isomorphism (due to Church) in order to make the criterion more precise. Four pages later, in an alleged 'proof' of his claim that a Suppes reduction obtains whenever a standard (Nagel—Quine) reduction does, he goes mysteriously astray by applying different notions of isomorphism and Suppes-reduction than those given earlier. Schaffner seems not to be aware that isomorphism is defined for models of the same similarity type or language, whereas in cases of scientific reduction it is customary to assume that the models (and theories) being related may be associated with different languages.

In itself, Schaffner's claim is not completely without foundation because — as I remarked earlier — Suppes' criterion *can* be interpreted as a semantic variant of the customary, derivational model of reduction; and this is probably how Suppes intends it to be read. But, in the absence of additional asusmptions, the claim cannot be formally proven. Nor does it seem to provide grounds, as Schaffner supposes, for a criticism of the Suppes concept.

7. See Theorems 4 and 5 of Eberle (1971). The translation in question is defined by what Eberle calls a *representing function*.

8. For more details on the 'justification' for treating only this fragment of the structuralist reduction concept, see Pearce (1982b). In particular, it is explained there why the notion of 'constraint' need not be explicitly incorporated in the present context.

9. Here and below, when $\rho \subseteq \text{Str}(\tau') \times \text{Str}(\tau)$ is an algebraic relation, I shall sometimes write $\rho(\mathfrak{M}', \mathfrak{M})$ instead of $(\mathfrak{M}', \mathfrak{M}) \in \rho$. And when, as in Sneed's case, ρ is single-valued, I shall sometimes re-express it as an *operation* $\rho: \text{Str}(\tau') \to \text{Str}(\tau)$, on the understanding that $\rho(\mathfrak{M}') = \mathfrak{M}$ iff $(\mathfrak{M}', \mathfrak{M}) \in \rho$. Algebraic operations on structures were first introduced by Gaifman (1974).

10. Regularity of a logic is explained in Feferman (1974). I shall also assume below that L satisfies the requisite Boolean closure properties needed for applying Feferman's uniform reduction theorem. I shall not trouble to distinguish here between projective and δ-projective classes of models, i.e. between the cases where S above is a single sentence or a set of sentences. In the latter case, assume that δ is a *compactness property* for L, in Feferman's sense.

11. Once again, see Feferman (1974) for the general concept of 'logic' intended here, and for specific examples of logics. Feferman's term for 'logic' is "model-theoretic language".

12. Henceforth the symbol ρ always denotes such a relation (or operation).

13. For later reference, I give a precise definition of this notion here. Let $\tau = \text{Sort}(\tau) \cup \text{Symb}(\tau)$ be a similarity type. A renaming of τ is then characterised by a bijective mapping $*$ of τ onto τ^* (τ^* a type), such that $\text{Sort}(\tau)$ is bijectively mapped onto $\text{Sort}(\tau^*)$, and each symbol R of τ is mapped to a symbol R^* of the same kind. $*$ induces a relation $\equiv^*$ between τ- structures and τ^*-structures with the property that $\mathfrak{M} \equiv^* \mathfrak{M}^*$ iff $\mathfrak{M}(R) = \mathfrak{M}^*(R^*)$ for each $R \in \text{Symb}(\tau)$. A logic L is said to admit name-changes if, for any appropriate type τ, and renaming $*$ of τ, $*$ induces a one—one mapping $*$ of $\text{Sent}(\tau)$ onto $\text{Sent}(\tau^*)$ with the property that if $\mathfrak{M} \in \text{Str}_L(\tau)$, $\mathfrak{M}^* \in \text{Str}(\tau^*)$ and $\mathfrak{M} \equiv^* \mathfrak{M}^*$, then $\mathfrak{M}^* \in \text{Str}_L(\tau^*)$ and $\mathfrak{M} \vDash_L \varphi \Leftrightarrow \mathfrak{M}^* \vDash_L \varphi^*$. (Cf. Feferman, 1974).

14. For any logic L, let Typ_L be the collection of types *admitted* by L. Let $\tau \in \text{Typ}_L$. $K \subseteq \text{Str}(\tau)$ is said to be an *L-elementary class* (in symbols: $K \in \text{EC}_L(\tau)$) iff K is the class of all L-structures of type τ that are models for some L-sentence φ of type τ. K is said to be *L-projective* ($K \in \text{PC}_L(\tau)$) iff there is a type $\tau^* \supseteq \tau$ and class $K^* \subseteq \text{Str}(\tau^*)$ such that $K^* \in \text{EC}_L(\tau^*)$ and $K^* \restriction \tau = K$, where $K^* \restriction \tau$ denotes the class of all restrictions of members of K^* to the subtype τ. A logic L is then said to have the *Interpolation Property* iff for each $\tau \in \text{Typ}_L$ and $K_1, K_2 \in \text{PC}_L(\tau)$.

$$K_1 \cap K_2 = 0 \Rightarrow \exists K \in \text{EC}_L(\tau)[K_1 \subseteq K \ \& \ K \cap K_2 = 0].$$

I shall sometimes drop the suffix L below, when the context is clear.

15. For more details on the above logics and the results available on interpolation, see Makowsky *et al.* (1976). For results and references on interpolation in modal, many-valued and intuitionistic logics, see Maksimova (1982).

16. The fact that here L and L' might be different, and that in general no

unique language can be associated with a theory, need not prejudice one's chances for establishing commensurability. For instance, two theories T and T' might not be first-order axiomatisable, yet there could be a perfectly reasonable translation between their first-order languages ($L_{\omega\omega}(\tau)$ and $L_{\omega\omega}(\tau')$). Likewise, one could try to relate T and T' in the case where these theories are normally associated with, or characterised by, quite different logics L and L' respectively; a situation dealt with in Chapter 4. One might then consider translations from Sent$_L(\tau)$ into Sent$_{L'}(\tau')$; or alternatively, translations from Sent$_{L*}(\tau)$ into Sent$_{L*}(\tau')$, for a suitable L^* in which a model-theoretic reduction can be defined.

NOTES TO CHAPTER 2

1. In what follows I concentrate mainly on Balzer's (1985b), but most of the points apply equally to Stegmüller's discussion in his (1986). One important difference should be noted, however. It is that Stegmüller, unlike Balzer, does not appear to question the *premises* ((1.1)—(1.4) above) of my argument. Thus he seems to agree that a reduction relation in his sense will generally embody a translation of a certain sort, which he calls an *abstract* translation (1986, Ch. 10). His chief criticism is that a translation of this kind will not generally preserve meaning (reference) and will not therefore ensure the commensurability of the reduced and reducing theories.

Balzer's analysis of commensurability in his (1985a) is intended to be somewhat broader and more wide-ranging than in his (1985b). But I find it hard to offer an accurate and sympathetic appraisal of the former work, because his explication there is based on an unanalysed notion of empirical system whose meaning is unclear to me. By contrast, his (1985b) reconstruction employs standard, model-theoretic concepts that are readily accessible, and it is directed specifically at the kind of argument I have given in Chapter 1.

2. Actually, Balzer (in D3 of his 1985b) defines the translation as a mapping of *formulas*, rather than only (closed) sentences; but there seem to be some technical difficulties with his formulation. For instance, since he permits the free variables of a formula to alter under translation, it follows from one of his conditions (D3(3)) that a quantified sentence might be translated by an open formula. And I think one may safely assume that this is an unintended consequence of the definition. In any case, the effect of his explication is to require for commensurability a literal translation of sentences; and there seems to be little loss of generality if we restrict our attention here to translation maps defined on sentences.

3. I own this last point to a remark of Haim Gaifman.

4. For the definitions involved here, see Chapter 1, note 13. Thus, in particular, the map ρ^* is defined by $\rho^*(\mathfrak{M}) = \mathfrak{M}' \Leftrightarrow \rho(\mathfrak{M}) \equiv^* \mathfrak{M}'$. From the definitions and the usual argument, one then obtains: $\rho^*(\mathfrak{M}) \vDash \varphi^* \Leftrightarrow \mathfrak{M} \vDash \Gamma^*(\varphi^*) \Leftrightarrow \mathfrak{M} \vDash \Gamma(\varphi)$, for appropriate models $\mathfrak{M}$ and sentences φ.

5. See for instance Krantz *et al.* (1971).

6. This point is discussed at length by Adams (1974).

7. See Narens (1974).

8. For examples of this type of approach to space—time theory, see Field (1980), Manders (1982), Mundy (1983) and Malhas (1984).

9. Representation theorems of this kind are proved in Mundy (1983).

10. Oikkonen's languages $M_{\kappa\kappa}$ are in fact 'approximations' of the languages $N_{\kappa\kappa}$ introduced by Hintikka and Rantala (1976). What Oikkonen proves, in particular, is that $M_{\kappa\kappa}$ characterises the so-called Δ-closure of $L_{\kappa\kappa}$ (for κ a limit of strongly inaccessible cardinals), and that it satisfies an Interpolation theorem. The set-theoretic assumptions associated with logics of this kind are probably no stronger than those required by the metatheory of Stegmüller's and Balzer's structuralist framework. In fact, Rantala (1980) remarks that a set theory admitting inaccessible cardinals would be appropriate for handling the sorts of power-class operations that figure prominently in the structuralist view of theories.

11. This condition is plainly appropriate only in the very special circumstances that $\mathfrak{M}'$ and $\rho(\mathfrak{M}')$ share the same domain. It is therefore not normally applicable in cases of reduction between scientific theories, though it can readily be adapted to cover the more general case. Relevant here are the generalised concepts of definability and interpretation discussed later in this chapter (cf. note 14 below), and mentioned again in Chapter 6. Kamp also remarks that this restrictive assumption can be relaxed within Montague's and his own approaches to formal language translation, but in his (1978) he does not carry out the generalisation in detail.

12. Actually, Balzer does not directly raise this point as an objection to my argument. But he does apparently require translations to be defined through the usual inductive clauses. See, once again, definition D3(3) of his (1985b).

13. If no special restrictions are placed on the logics involved, then definability in the logical sense will usually be taken to be a weaker notion than constructibility. But it is quite possible that Suppes would also admit set-theoretic constructions that are not explicitly defined in some chosen extension of first-order logic.

14. Since Gaifman's defining schemas represent an important generalisation of Beth-definability, it is worthwhile giving a brief account of them here. This will also serve us well as an illustration of how recursive translations can be defined between theories in situations where there is a *prima facie* 'difference

of domain'; in other words, where the correlated τ' and τ-models, $\mathfrak{M}'$ and $\mathfrak{M}$, have different universes. For cases of this kind we can then speak of the translation "preserving reference" in a generalised sense. To make the notation slightly less cumbersome, let us consider the case where $L(\tau)$, $L(\tau')$, etc., are countable, single-sorted, first-order languages. We can then follow the somewhat simpler treatment of Hodges (1975) and Pillay (1977), based on an unpublished paper by Gaifman. For the many-sorted version, see Gaifman (1974).

Retaining the notation of the text, let $P(v)$ be a one-place predicate of τ^*, not in τ', and for any τ^*-structure $\mathfrak{M}^*$, denote by $\mathfrak{M}^{*P}$ the substructure of $\mathfrak{M}^*$ whose universe is $\{x: \mathfrak{M}^* \vDash P(x)\}$. A Gaifman operation G then associates τ'-structures $\mathfrak{M}' = \mathfrak{M}^{*P} \upharpoonright \tau'$ with τ^*-structures $G(\mathfrak{M}') = \mathfrak{M}^*$, and the range of the operation is to be defined by an $L(\tau^*)$-theory, S. A defining schema D in Gaifman's sense then consists of the following:

(i)   a function which associates with each k-place predicate $R(v_1, \ldots, v_k)$ of τ^* (including equality) an $L(\tau')$-formula $\theta_R(\mathbf{x}_1, \ldots, \mathbf{x}_k)$, where $\mathbf{x}_i = x_{i,1}, \ldots, x_{i,n}$, for n a fixed positive integer;

(ii)   a fixed $L(\tau')$-formula $\sigma(\mathbf{u}, v)$, where $\mathbf{u} = u_1, \ldots, u_n$.

A τ^*-structure $\mathfrak{M}^*$ is then said to be *defined* in a τ'-structure $\mathfrak{M}' = \mathfrak{M}^{*P} \upharpoonright \tau'$ by D if there is a function f over n-tuples of elements of $\mathfrak{M}'$ such that

(a)   the universe of $\mathfrak{M}^*$ is $\{f(x_1, \ldots, x_n): x_1, \ldots, x_n \in \mathfrak{M}'\}$;

(b)   for all τ^*-predicates (including equality) R,
       $\mathfrak{M}^* \vDash R(f(\mathbf{x}_1), \ldots, f(\mathbf{x}_k)) \Leftrightarrow \mathfrak{M}' \vDash \theta_R(\mathbf{x}_1, \ldots, \mathbf{x}_k)$;

(c)   for all $y \in \mathfrak{M}', f(\mathbf{x}) = y \Leftrightarrow \mathfrak{M}' \vDash \sigma(\mathbf{x}, y)$.

Thus, the universe of $\mathfrak{M}^*$ is exactly determined by (the range of) the f-mapping (condition (a)); each predicate in the language of $\mathfrak{M}^*$ is in a generalised sense explicitly defined by a τ'-formula (condition (b)); and the map f itself is definable in $\mathfrak{M}'$ by a formula σ in the type τ' (condition (c)).

Clearly, when G is a rigid operation as described in the text, the fact that by Gaifman's result $G(\mathfrak{M}')$ is defined in $\mathfrak{M}'$ by D (for D a suitable defining schema) means that there is a recursive translation from $L(\tau^*)$-formulas into $L(\tau')$-formulas which respects G. Moreover, if G is determined by some reduction relation ρ, it follows that this translation induces a recursive translation from $L(\tau)$-formulas into $L(\tau')$-formulas which respects ρ.

15. Shelah's result is mentioned in Hodges (1975). Pillay (1977) also proves a number of relevant results about Gaifman operations. He has later extended Gaifman's theorem on definability (by removing the rigidity constraint) in a recent paper (1982).

16. The reader should consult Sette and Szczerba's paper, as well as

Szczerba (1977), for further details. Sette and Szczerba also provide algebraic characterisations of different types of interpretability in their (1983).

17. In his (1980), Hodges also offers an explicit characterisation of "*concrete* construction" and relates this notion to his word-constructions.

18. This last consequence of Balzer's definition could be avoided if only *some* common predicates R were required to be R-ρ-commensurable; in other words, if the universal quantifier of (0)(iii) were replaced by an existential one. But even this requirement — though more reasonable — seems much too strong in practice, because it does not allow for the possibility that models $\mathfrak{M}$ and $\rho(\mathfrak{M})$ might have disjoint domains (nor, of course, that a predicate R simply has a different extension in $\mathfrak{M}$ than in $\rho(\mathfrak{M})$). Moreover, it is hard to concur with Stegmüller's suggestion (see his 1986. p. 305) that condition (0)(iii) is required to ensure that a translation preserves (the extensional component of) meaning (is *bedeutungerhaltend*). For the reasons given, (0)(iii) is clearly not a necessary condition for reference-preservation. Moreover, even if coupled with the requirement (0)(ii) of *literalness*, it would hardly be a sufficient condition either; for this one would need the denotation of R to be the same in $\mathfrak{M}'$ and $\mathfrak{M}$, for *all* pairs $\langle \mathfrak{M}', \mathfrak{M} \rangle \in \rho$, not merely some pairs as (0)(iii) requires.

NOTES TO CHAPTER 3

1. Laudan does not specifically reject the semantic conception of truth. But he does claim that truth is not a methodological goal of science and has no bearing on evaluating the cognitive content of a theory or its rational acceptability.

2. Another example of the conventionalist flavour of Laudan's approach is afforded by his treatment of ad hocness, discussed below. In particular, Laudan allows what Popper calls "conventionalist strategems in science" to be potentially progressive.

3. See especially Laudan (1981b) and (1984).

4. See, for example, Laudan (1980).

5. I shall·be concerned here chiefly with the views Laudan expounds in his (1977). I can find no evidence from Laudan's later writings that he has changed his opinions on any of the points that I shall be treating below.

6. As I remarked in Chapter 1, Kuhn himself strongly denies the charge that his conception of science is irrationalist. And at least in *The Structure of Scientific Revolutions* (to which Laudan refers), Kuhn's characterisation of incommensurability is considerably more complex and subtle than the reformulation that Laudan offers.

7. Feyerabend summarises his own view of incommensurability and its consequences, and contrasts this with Kuhn's account, in Feyerabend (1977).

8. It is not entirely clear whether Laudan's own illustrations of shared problems (cf. his 1977, pp. 143—145) all enter into the same category. For instance, the "problem of free fall", said to be a common concern for both eighteenth-century Newtonians and Cartesians, might be regarded as being of type (i) or of type (ii), depending on how it is characterised. A much better example from Laudan's point of view is the case of the problem of light reflection (as dealt with by late seventeenth-century optics). His claim that this problem was characterisable independently of the various theories competing at the time is no doubt grounded on the idea that the basic laws of geometrical optics were shared by both the wave and corpuscular theories of light and were thus 'nontheoretical' with respect to those theories. But it is evidently a non-trivial matter to identify an acceptable theory (or fragment) of geometrical optics and to establish its status in widely differing theories of light.

9. It is also possible to read Laudan as intending this second claim to hold for only some, rather than all, terms in S. However, this would not alter significantly the substance of my remarks below.

10. Some commentators have indeed criticised Laudan on just this point. See, for example, Gutting (1980).

11. In general, one might allow for some differences in methodological standards of appraisal, accuracy, and so forth, providing that these standards are rationally comparable in a suitable sense. In more recent works, especially his (1984), Laudan discusses such 'higher' level methodological differences between research traditions at some length. But the analysis he provides does not protect his model of progress from the difficulties raised here, because even shared cognitive norms will be of little consolation if there is genuine incommensurability at the 'ground' level, i.e. at the level of concepts and problems.

12. Actually, Laudan lists these three factors as criteria of *choice*, but it seems likely that the context of pursuit is the more appropriate one in the present case of a freshly emerging research tradition; and the same factors should apply here as well.

13. It would be tempting to argue that mathematical research traditions share some similar normative standards of simplicity, coherence, consistency, and so forth; but this is by no means obvious. For example, *consistency* of theories is a major motivation in the construction of intuitionist mathematics, but is only one among many other desiderata within classical mathematics. And in the modern research tradition which seeks to reconstruct dialectical and paraconsistent logic and mathematics, localised *inconsistency* is an accepted property of theories.

14. Naturally, if the research traditions are completely incommensurable in that they share *no* empirical problems, they cannot possess anomalies in Laudan's sense of the term; they may contain anomalies in the more traditional sense, however. The notion of 'empirical' problem is clearly to be understood here in a much wider-than-usual sense. This is in keeping with Laudan's own usage, according to which empirical problems may relate to a domain of abstract entities, just as they may be concerned with, say, 'observable' events and processes. Likewise, for Laudan, research traditions are not confined to the natural, or even the 'empirical', sciences (in the narrower sense).

NOTES TO CHAPTER 4

1. Though his 1949 paper is generally reckoned as the starting point of modern systematic studies of reduction, Nagel's work on reduction in fact goes back a long way before that date; see, for example, his (1935) and (1936).

2. General overviews of our model of reduction can be found in Pearce and Rantala (1983c, d, e).

3. There are, of course, exceptions. For instance, Erhard Scheibe (especially in his 1982 and 1983) draws a distinction between different reduction types that is similar to the one made here. What he terms the *potential explanation* of one theory by another corresponds roughly to theoretical reduction in the sense just described, whilst his notion of *factual explanation* as an intertheoretic relation seems quite close to an explanatory reduction in my sense. Scheibe's analysis of these two types of reduction differs somewhat from the one given below, however, chiefly in that he employs a quite different formal framework; the precise connections between our two approaches remain to be explored.

The difference between theoretical and explanatory reduction is also sometimes emphasised in the literaure on approximative reduction in physics. One finds, for example, the distinction between 'the approximation of laws' and 'the approximation of solutions'; see, e.g., Schmidt (1984), Ehlers (1986).

4. An example of a well-developed structural approach to reduction, different from Sneed's, can be found in the work of G. Ludwig and his followers. See, for instance, Ludwig (1978) and various contributions in Hartkämper and Schmidt (1981).

5. In this quotation I have reversed Sneed's priming convention. Thus T is to be understood as the reduced and T' the reducing theory.

6. Sneed's and Mayr's accounts of reduction are contrasted, and their logical properties compared, in Pearce (1982b).

ROADS TO COMMENSURABILITY

7. For the notion of physical determinationism, see Hellman and Thompson (1975) and Tennant (1985a). Some logical aspects of the determinationist thesis (and its comparison with reductionism) are treated in Pearce (1985).

8. The models might, for instance, be of the classical Tarskian kind, or perhaps nonclassical structures of Kripke (or some other) sort.

9. Notice that this fact does not conflict with my claim in Chapter 2 that empirical theories do not as a rule embody fixedly interpreted mathematical systems. The point is rather this: that an empirical theory may appeal *loosely* to some underlying mathematical system (e.g. of real numbers), or it may make *explicit* use of a certain well-defined mathematical structure (Measure Space, Hilbert Space, etc.). In either event, the mathematics is usually informally construed, and a definite semantic interpretation or formalisation will be supplied, if at all, only at the stage of logical reconstruction.

10. Hempel (1965) calls this *approximative* explanation. Views, and terminologies, similar to Hempel's, have been expressed by many writers including Putnam (1965) and Nagel (1970). For comments on and criticism of them, see Pearce and Rantala (1984b; 1985; 1988).

11. The restriction to model-theoretic correlations that are *functional* need not be rigidly adhered to. We could equally well consider definable (many-valued) *relations* between structures (see below). On the other hand, the generalisation to *translations* that are *partial* (i.e. not everywhere defined) is one that I shall not consider here; see, however, Chapter 7, §2.

12. The possibility that no nontrivial translation is involved is, of course, taken care of by the fact that identity or literal translations are included as a special case (see below).

13. A more detailed account and defence of the following conception of 'theory' may be found in Pearce and Rantala (1983a).

14. Typically the classes N and M will be closed under isomorphism.

15. The possible formal connections holding between different *symmetries, R and R'*, are explored in Pearce and Rantala (1983b; 1988).

16. If σ' is an L'-sentence axiomatising the domain K' of the mapping F, then a potential explanation of φ in the framework of T' might have the form:

$$\sigma', I(\theta_1), \ldots, I(\theta_n), I(\psi_1), \ldots, I(\psi_k) \vDash_{L'} I(\varphi).$$

This is not the only possibility, of course, since T' might also entail the translation of φ under a different set of initial conditions and auxiliary assumptions.

17. For a precise characterisation of limiting case correspondence, see Pearce and Rantala (1984a; 1988).

18. For the following relations, assume that the logics L and L' involved are the same.

19. Though I shall not introduce them here, it is clear that the α' branch of Figure 1 can be extended by adding approximative interpretations, embeddings, etc.

20. Cf. note 14 above.

21. Though Hoering presents an interesting discussion of the problem, his objections to some of the standard accounts of reduction do not strike me as entirely compelling. This applies in particular to his suggestion that the existence of a reduction in the Sneed sense might essentially depend on the cardinalities of the collections of models of the relevant theories. The problem here is that in the normal way such collections will not be sets (in the sense of, say, ZF set theory), and therefore not possess cardinalities. (I owe this remark to Veikko Rantala; see also his 1980.) I discuss Hoering's approach at greater length in my (1983).

22. Even in such a paradigmatic example of reduction as this, the logical derivation may require approximating or idealising assumptions (e.g. gas molecules represented as mass points, perfect elasticity of collisions, etc.). For this reason, one cannot usually assume that the additional premises required for reduction are literally true, or even well confirmed; their epistemic acceptability has to be gauged on other grounds.

23. Naturally, the two views just distinguished represent very different positions with regard to the mind—body problem. The point I wish to emphasise here is that the psychophysical identity thesis is not always taken to be an essential ingredient of psychophysical reductionism, and some philosophers would even regard it as logically independent of the reductionist thesis. In this case, therefore, the extent to which 'real systems' can be identified outside the context of a particular theory or pair of theories, and the extent to which this identification has to play a central role in characterising reduction, are by no means trivial matters to settle.

24. This is in effect the form which Searle's argument against strong AI takes. Put in very crude terms, his claim is that since mental events have semantic content and computer programs are by definition purely syntactical, it follows from the nonreducibility of semantics to syntax that no reduction of the kind required by strong AI is possible. Naturally, the crucial premise that semantics cannot be reduced to syntax is here taken to be a relatively unproblematic assumption acceptable to both parties in the debate.

25. I am suggesting, in particular, that one should be able to analyse and evaluate 'borderline' and controversial instances of reduction on the basis of these kinds of formally describable features. For this, of course, one needs to have precisely reconstructed examples. One case of this type that merits further investigation is an example given by Balzer (1984) of a purported reduction of classical to relativistic kinematics. The unexpected feature of this example is

that the relation Balzer defines is 'exact' rather than 'approximate' (as one would usually assume of any reduction between these theories). However, reasonable grounds for questioning the adequacy of the reduction are in this case very clearly expressible in formal terms. First, according to Balzer, there is no definable translation from the 'reduced' to the 'reducing' theory; (a claim which, however, would seem to require further justification). And, secondly, the structural correlation does not apparently commute (in a suitable sense) with the usual Galilean and Lorentz transformations; whereas commutativity of some kind is a natural desideratum here.

26. For a discussion of the role of reduction and other intertheory relations as determinants of scientific research, see Krüger (1980).

NOTES TO CHAPTER 5

1. I shall not enter into further details here, except to remark that many of the difficulties with the structuralist account seem to flow directly from those properties of theory nets and evolutions that I pointed out in Chapter 1. For a constructive and particularly penetrating critique (of a different sort) of the structuralist theory dynamics, see Niiniluoto (1984, Ch. 6).

2. For earlier attempts to deal with some of the ideas and themes developed in this chapter, see Pearce and Rantala (1983d) and (1984c).

3. See Hintikka (1981; 1984).

4. For a discussion of explanation-seeking (or 'es') questions and their relevance for Laudan's treatment of empirical problems, see Sintonen (1984). There Sintonen also provides a detailed examination of Laudan's problem-solving model within a structuralist type of framework. In keeping with the Sneedian perspective, Sintonen construes *problems* as *structures* or *classes of structures*. Formally speaking, this is a perfectly legitimate move. But there are several reasons. I believe, to prefer something closer to the usual statement view of problems. Here are three. First, the syntactic representation (1) of the problem-solving process brings out nicely the formal analogies with deductive-nomological types of explanation. This is reasonable if one supposes (as Sintonen does) that empirical problems are largely of the 'es' type. Secondly, a considerable burden of extra work would be needed to show why different structures (in a given class) intuitively represent the same problem or problem-type. The linguistic formulation usually makes this clear at once. And, finally, the main tool I shall use for handling the reappearance of a given problem within different conceptual frameworks is the method of *translation*. Again, this applies most naturally to statements, or propositions in statement form.

5. Other desiderata here would be to make the domain of theories as large as possible and to include in it, where feasible, the most important or central

theories of the research tradition. Evidently these different constraints need not be jointly realisable, or at least they need not lead to a unique representation of the tradition. Different structures might therefore be chosen (perhaps for different purposes) as suitable candidates.

6. Cf. Pearce and Rantala (1983d).

7. Such 'generalised' structures can, in their turn, also be conceived as ensembles.

8. Actually, in the paper referred to, the term 'theory ensemble' does not appear. For expository reasons we borrowed from structuralist terminology, and employed a generalisation of their concept of net. The notion of ensemble can readily be applied there, however.

9. This list is not intended to be exhaustive. But it illustrates how several of the most important types of methodological constraints on theory development might receive a rather natural and exact reformulation in the context of theory ensembles.

10. For an explicit treatment of symmetries in intertheory relations, see Pearce and Rantala (1983b; 1988). In particular, in the model-theoretic setting one can give a precise reconstruction of symmetry-preservation and of relations between different symmetries.

11. I would not insist that this condition be unfailingly met. But it does facilitate the definition of *uniform* relations between ensembles, as discussed in the next two paragraphs.

12. Readers of Pearce and Rantala (1984c) should note that a certain ambiguity arises in that paper. A claim similar to (4) is asserted ((7) on p. 396) which requires the premise

$$M_1 \restriction \tau = \mathrm{Mod}_L \, I_1(\theta_1),$$

though this identity is only *implicitly* assumed in the paper. In case one is able to infer only the weaker inclusion

$$M_1 \restriction \tau \subset \mathrm{Mod}_L \, I_1(\theta_1),$$

claim (4) should be modified to

$$(4') \qquad \sigma, I_1(\theta_1), I_1(\psi) \vDash_L I_1(\varphi);$$

where σ is an $L(\tau)$-sentence axiomatising $M_1 \restriction \tau$.

13. Statement (9) of Pearce and Rantala (1984c, p. 397) should be corrected to conform to the pattern of (5) above. In case the problem-solving situation is represented by $(4')$ rather than (4) (see note 12 above), clause (5) should be restated in the slightly more complex form:

$$(5') \qquad \sigma', I'(\sigma), I(\theta_1), I(\psi) \vDash_{L'} I(\varphi).$$

This still yields a representation of the problem-solving scheme (3) within the framework of **T**′.

14. This assumption is, of course, important here. For, the connections between ensembles based on other kinds of orderings might be quite different; cf. Pearce and Rantala (1984c).

15. This need not imply that T' absorbs a part of T *unaltered*. As in the example from mechanics mentioned below, T's success might be seen (from T') to be not only restricted in scope but also of merely *approximate* validity, and K' may be defined by suitable *limit* conditions reflecting this.

16. Sintonen (1984) (discussing in the context of theory nets) seems to have an analogous idea in mind when he suggests (p. 142) that the most important problems are those associated with theories that are *close to the base*.

NOTES TO CHAPTER 6

1. For a more exact formulation of the paradox of meaning variance, and a detailed analysis of it, see Giedymin (1970). I discuss briefly my own approach to a solution of the paradox at the end of Chapter 7.

2. Frank's account can be found in his (1946). It was recalled and discussed in the context of the debate over incommensurable theories by Giedymin (1973).

3. Sneed (1971), for example, provides a number of arguments to support the view that the term 'mass' is theoretical in the context of classical mechanics.

4. Detailed defence of the claim that classical and relativistic mechanics deal with the same concept of mass is brought, for example, in Krajewski (1977) and Yoshida (1977). However, Krajewski from the outset appears to rule out the need for any semantical analysis of terms like 'mass'. His otherwise thorough treatment of the correspondence relation is cast in exclusively syntactic form, and this makes it hard to evaluate the problem of meaning change on the basis of his model. Yoshida, like several critics of the meaning variance thesis, argues that properties like *conservation* and *velocity independence* are not intrinsic to the meaning of classical 'mass'. However, along with most disputants, including Kuhn and Feyerabend, Yoshida provides no meaning theory precise enough to judge the matter impartially. His 'counterexamples' to Kuhn are likewise only briefly and inconclusively sketched. Moreover they are directed at rather extreme positions — e.g. that mass conservation is a non-refutable (meaning) postulate of CM — which do not appear to be necessary premises of the argument for meaning change. Both Krajewski and Yoshida seem to think that their own characterisation of mass as "that property of a body in virtue of which it resists a change in motion" is sufficiently clear and precise to require no further analysis, besides being obviously independent (in meaning) of one's commitment to any particular theoretical framework (e.g. the classical or relativistic). I confess to finding both assumptions questionable.

5. See especially Scheffler (1967), Putnam (1973; 1975a).

6. This line of argument against the efficacy of the causal theory in this context is cogently developed by Fine (1975). In the same paper, Fine also makes an interesting, constructive contribution to the problem of meaning and reference change. It is possible that Fine's approach shares some broad similarities with the sort of position that I adopt here and especially in the last half of the next chapter. Unfortunately, his views are only presented in the form of a brief and informal sketch, so I am unable to present any conclusive comparison here.

7. The seminal works on this theory are Dummett (1973), Prawitz (1977), Tennant (1987) and various essays in Dummett (1978). There is already a vast and growing secondary literature; see Tennant (1987) for references.

8. Tennant (1985b; 1987) shows that intuitionistic and even minimal relevant logic are adequate for 'Popperian science' in the sense that contradictions between theories and observation reports that are expressible in classical first-order logic are provably contradictory in the weaker logics too. This is a step in the direction of applying the nonrealist semantics to aspects of theory testing in science. It does not, of course, supply a fully fledged meaning theory for empirical science; nor would the weaker logics discussed by Tennant normally be adequate for reconstructing physical theories, in the sense of adequacy defined here (Chapter 4).

9. Putnam has produced numerous variations of the main argument in different essays and books; e.g. Putnam (1978; 1980; 1981).

10. This remark is developed into a detailed criticism of Putnam's argument in Pearce and Rantala (1982a, 1982b).

11. The geometry-arithmetic relation is of course a classic example of reduction which falls under the various schemas of interpretability cited in Chapter 2 above. The case of rigid-body and particle mechanics is the paradigm example of structuralist reduction analysed by Adams (1959) and Sneed (1971); some aspects of it are discussed on a more logically oriented footing in Pearce (1982b). For an example of syntactically and semantically analysed translation between intuitionistic and classical logic, see Smorynski (1975).

12. There is no need to enter into further details here. However, the generalised notions of definability and interpretability mentioned in Chapter 2 indicate the main lines along which translatability of this kind can be characterised. There are, of course, also examples of translation where even reference preservation in the 'sophisticated' sense is not a desideratum (e.g. when preservation of self-reference is required); see Burge (1978).

13. Przełęcki's class M of *intended* models should not be confused with the class of *all* models of a theory, also denoted by M in earlier chapters. For Przełęcki models in M pick out the 'actual' referents of terms, rather than the extensions which are assigned by arbitrary models of the theory. In fact, elements of Przełęcki's class M need not even be in the customary sense

models of the theory's laws. Thus, his notion of *truth* does not coincide with "being a logical consequence of the theory".

14. A similar type of argument against Field can be found in Papineau (1979, pp. 153—156).

15. In the more general case, the models $\mathfrak{M}$ and $\mathfrak{M}'$ are not required to have identical domains, in which case the interpretations they assign to t must overlap or be otherwise related in an appropriate way.

16. Under certain, rather strong, assumptions Carnap's approach to the semantics of empirical theories can be shown to lead to a situation of non-translatability between the theoretical expressions of two mutually inconsistent theories; see Williams (1973). Przełęcki apparently accepts Williams' analysis of this feature, and incorporates it in his (1979).

17. At one point, Papineau (1979, p. 152) defends Field by reminding his reader that Field is primarily dealing with the 'external' reference of terms, say from CM, rather than the *relation* those terms bear to concepts from a later theory, e.g. RM. Papineau assumes that the epistemic question of choosing between CM and RM is already settled in favour of the latter theory. But it should be noted that Field expressly intends his approach to provide an argument against the incommensurability thesis. For this it would seem he requires a means of *comparing* the referents of terms from CM and RM, in a way which can be seen to contribute to the rational appraisal of the two theories.

18. I think it also unlikely that Field can be regarded as a *reformer* of semantic theory. As far as I can tell he introduces only one minor and unnecessary modification, that the notion of a *structure* is defined in the context of *sentences* rather than the total vocabulary of the language (p. 477). In other respects his truth definition is exactly like that of ordinary referential semantics, where partial denotation is substituted for denotation.

19. Hesse's interpretation does not seem to square with Field's own, however; see Field (1973, p. 475).

20. To be fair, Hesse is cognisant of some of the less attractive properties that this kind of translation would have. She writes, for instance, that "truth relations internal to one theory are not invariant to the translation process" (Hesse, 1980; p. 157). She does not, however, consider the possibility that these very features might even prevent one from formally *defining* the translation in an adequate way.

21. These remarks make it clear that, contrary to Levin's claim, his 'translation' cannot be a *recursive* mapping of expressions. In the first place, it is evidently not defined using the usual inductive clauses. Secondly, for the translation to be recursive, RM (or the current physical theory chosen for determining truth values) would have to be in the logical sense a *decidable* theory; this is, to say the least, highly implausible.

22. This point can be illustrated by a trivial example, analogous to one used by Levin. The French adjective *propre* divides neatly (apart from nuances of meaning) in English translation: into the words 'own' and 'clean'. The rule is extremely simple: one checks whether the French word comes *before* or *after* the qualified noun, and translates by one or other of the English words accordingly; inspection of the *French* sentence is all that is required. Plainly, Levin's proposal to translate Newtonian English, and in particular to divide 'mass' in translation, does not follow this kind of pattern. Thus, his own analogy with natural language translation breaks down.

23. Cf. Kuhn's remark, in the passage quoted above, implying that Newton's second law would have to be valid in any framework capable of sustaining an adequate translation of CM.

24. In Section 6 of his (1983), under the heading 'The Invariants of Translation', Kuhn discusses various properties that translation should preserve. Most of his remarks are couched in metaphorical terms, like 'lexical network', 'criterial linkage', 'multi-dimensional structures', and so on; any comparisons one might make between Kuhn's criteria of adequacy and conditions that one might reconstruct in the framework of formal semantics are therefore bound to be *highly* conjectural. All the same it is sometimes tempting to make such comparisons. For instance, one might draw an analogy between Kuhn's assertion that "different languages impose different structures on the world", and the property that different formal languages have different kinds of model-types as semantic structures. The latter feature occurs, of course, in examples of the kind mentioned in note 11 above, where theories may be based on different vocabularies and sorts, or even on different underlying logics. However, this analogy might be questioned because Kuhn holds that speakers of intertranslatable languages must share the same 'lexical structures', whereas any standard theory of formal language translation allows the semantic structures associated with the languages to be very different. To maintain the analogy one would therefore have to interpret Kuhn as imposing a very strong, and to my mind implausible, constraint on acceptable translation.

25. At least if one ignores the uninteresting 'domain' of stationary particles.

26. Some of these models of intertheory explanation are discussed in Pearce and Rantala (1985; 1988). I comment briefly on the Popperian model at the end of the next chapter.

NOTES TO CHAPTER 7

1. My (1979/85) is devoted primarily to translation between languages, based on the construction of common, conservative extensions of them in this sense.

2. On Poincaré's conventionalism, see Giedymin (1977). For a critique of Popper's critique of instrumentalism with special reference to Duhem and to medieval astronomy, see Giedymin (1976). These essays are reprinted (with some changes) as Chapter 1 and 3, respectively, of Giedymin (1982).

3. Cf. Giedymin (1982, p. 83). The reader is recommended to consult Giedymin's book on all matters related to Poincaré's epistemology and the conventionalist tradition. Here I can only provide a very fragmented sketch of some aspects of these issues which seem especially relevant to problems of translation and commensurability.

4. See, Giedymin (1980 and 1982, Ch. 2).

5. See, especially, Adjukiewicz (1934a) and (1934b) (English translations in Giedymin, 1978). For the connections between Poincaré's and Ajdukiewicz's conventionalism, see Giedymin (1978, Introduction) or (1982, Ch. 4).

6. Adjukiewicz himself regarded the claim that physical theories can be associated with closed, connected languages in his sense as an idealising assumption. Giedymin (1982, Ch. 4) suggests that doubts concerning the adequacy of this assumption may have contributed to Ajdukiewicz's own decision to abandon the stronger form of (radical) conventionalism (and retain only a more moderate version) after 1936.

7. One might also put the distinction in the following way: that in a closed framework every sentence is *conditionally* true-or-false, where the condition in question is itself factually meaningful.

8. Gaifman (1984) discusses, with the help of examples, this last feature (minimimal ontic commitment) in terms of the effectiveness, or computational complexity, of the mapping t. His approach to translation explicitly rules out as adequate, for instance, the kind of mapping which Levin (1979) tries to define (that I discuss in the last chapter). One of Gaifman's examples of an improper translation is in fact completely analogous to Levin's; where translation is defined using the concept of truth in one of the frameworks concerned.

9. Given the form of the reconstruction employed in Pearce and Rantala (1984a), a suitable choice for L would be the logic $L_\omega \omega$ (not $L_{\omega_1 \omega}$, as was suggested there — an error which Johan van Benthem spotted and kindly pointed out to us). However, it is likely that weaker logics would suffice for characterising the correspondence relation, if the form of the reconstruction were to be modified somewhat.

10. In a nonstandard model of the reals, every finite, nonstandard real r can be associated a unique standard real st(r), called the *standard part* of r, with the property that the difference between r and st(r) is infinitesimal. For the generalisation of the standard part operation to real-valued functions (and to more general collections), see Pearce and Rantala (1984a).

11. A little more precisely, the standard parts of u_1 and u_2 are set equal. For details of the translation sketched here, see Pearce and Rantala (1984a, pp. 73—76).

12. Since the auxiliary assumption which permits the derivation of CM from RM sets *limits* on the validity of CM, some readers would probably prefer (here and below) a more qualified formulation of the sort: "RM explains *the successful part of* CM". However, since most law-explanations in science set limits and restrictions on the validity of the explanandum, these two formulations reflect at most a pragmatically determined difference of degree. I shall stick to the apparently stronger formulation here, since it highlights better the differences between the present account of the CM/RM relation and alternatives of the sort discussed in this paragraph and the next.

Whilst I have been stressing in general the indicative, truth-functional character of the CM/RM relation, the model-theoretic analysis which supports this view can also be adapted to handle the idea that the logical relation of CM to RM may take the form of a counterfactual inference. For a precise reconstruction of the latter within a Lewis-style semantics for counterfactuals, see Pearce and Rantala (1988) and a forthcoming paper by Veikko Rantala: 'Scientific Change and Counterfactuals'.

13. The 'paradox' usually assumes that logical disjointedness of this kind is already a consequence of the meaning-variance assumption; but it seems to me that additional premises would be required to ensure this.

BIBLIOGRAPHY

Adams, E. W., 'The Foundations of Rigid Body Mechanics and the Derivation of its Laws from Those of Particle Mechanics', in L. Henkin, P. Suppes and A. Tarski (eds), *The Axiomatic Method*, North-Holland, Amsterdam, 1959.

Adams, E. W., 'Model-Theoretic Aspects of Fundamental Measurement', in L. Henkin (ed.), 1974.

Ajdukiewicz, K., 'Sprache und Sinn', *Erkenntnis* **4** (1934a), 100—138.

Ajdukiewicz, K., 'Das Weltbild und die Begriffsapparatur', *Erkenntnis* **4** (1934b), 259—287.

Balzer, W., 'Incommensurability and Reduction', in I. Niiniluoto and R. Tuomela (eds), 1979.

Balzer, W., 'On the Comparison of Classical and Special Relativistic Space—Time', in W. Balzer *et al.* (eds), 1984.

Balzer, W., 'Was ist Inkommensurabilität?', *Kant Studien* **76** (1985a), 196—213.

Balzer, W., 'Incommensurability, Reduction and Translation', *Erkenntnis* **23** (1985b), 255—267.

Balzer, W. and Moulines, C.-U., 'Die Grundstruktur der klassischen Partikelmechanik und ihre Spezialisierungen', *Z. Naturforsch.* **36A** (1981), 600—608.

Balzer, W., Pearce, D. and Schmidt, H.-J. (eds) *Reduction in Science: Structure, Examples, Philosophical Problems*, D. Reidel, Dordrecht, 1984.

Balzer, W. and Sneed, J. D., 'Generalized Net Structures of Empirical Theories, I and II', *Studia Logica* **36** (1977), 195—212; and *37*(1978), 167—194.

Benthem, J. van and Pearce, D., 'A Mathematical Characterisation of Interpretation between Theories', *Studia Logica* **43** (1984), 295—303.

Born, M., *Einstein's Theory of Relativity*, Dover, New York, 1982.

Burge, T., 'Self-reference and Translation', in F. Guenthner and M. Guenthner-Reutter (eds), 1978.

Craig, W., 'Three Uses of the Herbrand-Gentzen Theorem in relating Model Theory and Proof Theory', *J. Symbol. Logic* **22** (1957), 269—285.

Dummett, M., 'The Philosophical Basis of Intuitionistic Logic', in H. E. Rose and J. C. Shepherdson (eds), *Logic Colloquium '73*, North-Holland, Amsterdam, 1973; also reprinted in Dummett, 1978.

Dummett, M., *Truth and other Enigmas*, Harvard University Press, Cambridge, 1978.

Eberle, R. A., 'Replacing One Theory by Another under Preservation of a Given Feature', *Phil. Sci.* **38** (1971), 486—501.

Ehlers, J., 'On Limit Relations between, and Approximative Explanations of, Physical Theories', in R. Barcan Marcus *et al.* (eds), *Logic, Methodology and Philosophy of Science VII*, Elsevier, Amsterdam, 1986.

Feferman, S., 'Two Notes on Abstract Model Theory, I', *Fund. Math.* **82** (1974), 153—165.

Feyerabend, P. K., 'Explanation, Reduction and Empiricism', in H. Feigl and G. Maxwell (eds), *Scientific Explanation, Space, and Time*, University of Minnesota Press, Minneapolis, 1962.

Feyerabend, P. K., 'Consolations for the Specialist', in I. Lakatos and A. Musgrave (eds), *Criticism and the Growth of Knowledge*, CUP, Cambridge, 1970.

Feyerabend, P. K., 'Changing Patterns of Reconstruction', *Brit. J. Phil. Sci.* **28** (1977), 351—369.

Field, H. H., 'Theory Change and the Indeterminacy of Reference', *J. Philosophy* **70** (1973), 462—481.

Field, H. H., *Science without Numbers*, Blackwell, Oxford, 1980.

Fine, A., 'How to Compare Theories: Reference and Change', *Nous* **9** (1975), 17—32.

Frank, P., *Foundations of Physics*, Chicago University Press, Chicago, 1946.

Gaifman, H., 'Operations on Relational Structures, Functors and Classes, I', in L. Henkin (ed.), *Proc. of the Tarski Symposium, AMS Proc. Pure Math.* **25**, Providence, 1974.

Gaifman, H., 'Ontology and Conceptual Frameworks', Part I, *Erkenntnis* **9** (1975), 329—335; Part II, *Erkenntnis* **10** (1976), 21—85.

Gaifman, H., 'Why Language?', in W. Balzer *et al.* (eds), 1984.

Gärdenfors, P., 'A Pragmatic Approach to Explanations', *Philosophy of Science*, **47** (1980), 404—423.

Giedymin, J., 'The Paradox of Meaning Variance', *Brit. J. Phil. Sci.* **21** (1970), 257—268.

Giedymin, J., 'Logical Comparability and Conceptual Disparity between Newtonian and Relativistic Mechanics, *Brit. J. Phil. Sci.* **24** (1973), 270—276.

Giedymin, J., 'Instrumentalism and its Critique: a Reappraisal', in *Essays in Memory of I. Lakatos* (*Boston Studies in the Philosophy of Science* **39**), D. Reidel, Dordrecht, 1976.

Giedymin, J., 'On the Origin and Significance of Poincaré's Conventionalism, *Stud. Hist. Phil. Sci.* **8** (1977), 271—301.

Giedymin, J. (ed.), *Kazimierz Ajdukiewicz: The Scientific World-Perspective and other Essays, 1931—1963*, D. Reidel, Dordrecht, 1978.

Giedymin, J., 'Hamilton's Method in Geometrical Optics and Ramsey's View of

Theories', in D. H. Mellor (ed.), *Prospects for Pragmatism*, CUP, Cambridge, 1980.

Giedymin, J., *Science and Convention*, Pergamon Press, Oxford, 1982.

Guenthner, F. and Guenthner-Reutter, M. (eds), *Meaning and Translation*, Duckworth, London, 1978.

Gutting, G., 'Conceptual Structures and Scientific Change', *Stud. Hist. Phil. Sci.* **4** (1973), 209–230.

Gutting, G., 'Review of L. Laudan: *Progress and its Problems*', *Erkenntnis* **15** (1980), 91–103.

Hartkämper, A. and Schmidt, H.-J. (eds), *Structure and Approximation in Physical Theories*, Plenum Press, New York, 1981.

Hellman, G. and Thompson, F., 'Physicalism: Ontology, Determination and Reduction', *J. Phil.* **72** (1975), 551–564.

Hempel, C. G., *Aspects of Scientific Explanation*, Free Press, New York, 1965.

Henkin, L. (ed.), *Proc. of the Tarski Symposium, AMS Proc. Pure Math.* **25**, Providence, 1974.

Hesse, M., 'Truth and the Growth of Scientific Knowledge', in F. Suppe and P. Asquith (eds), *PSA 1976*, Philosophy of Science Assoc., East Lansing, 1977; also reprinted in M. Hesse, 1980.

Hesse, M., *Revolutions and Reconstructions in the Philosophy of Science*, Harvester Press, Brighton, 1980.

Hintikka, J., 'On the Logic of an Interrogative Model of Scientific Inquiry', *Synthese* **47** (1981), 69–83.

Hintikka, J., 'The Logic of Science as a Model-Oriented Logic', in P. Asquith and P. Kitcher (eds), *PSA 1984*, Philosophy of Science Assoc., East Lansing, 1984.

Hintikka, J. and Rantala, V., 'A New Approach to Infinitary Languages', *Ann. Math. Log.* **10** (1976), 95–115.

Hodges, W., 'A Normal Form for Algebraic Constructions, II', *Logique et Analyse* **71–72** (1975), 429–480.

Hodges, W., 'Constructing Pure Injective Hulls', *J. Symbol. Logic* **45** (1980), 544–548.

Hoering, W., 'Anomalies of Reduction', in W. Balzer *et al.* (eds), 1984.

Kamp, H., 'The Adequacy of Translation between Formal and Natural Languages', in F. Guenthner and M. Guenthner-Reutter (eds), 1978.

Krajewski, W., *Correspondence Principle and the Growth of Science*, D. Reidel, Dordrecht, 1977.

Krantz, D. H., Luce, R. D., Suppes, P. and Tversky, A., *Foundations of Measurement*, Vol. 1, Academic Press, New York, 1971.

Krüger, L., 'Intertheory Relations as a Tool for the Rational Reconstruction of Scientific Development', *Stud. Hist. Phil. Sci.* **11** (1980), 89–101.

Kuhn, T. S., *The Structure of Scientific Revolutions*, University of Chicago Press, Chicago, 1962.

Kuhn, T. S., 'Commensurability, Comparability, Communicability', in P. Asquith and T. Nickles (eds), *PSA 1982*, Philosophy of Science Assoc., East Lansing, 1983.

Laudan, L., *Progress and its Problems*, University of California Press, Berkeley, Los Angeles, London, 1977.

Laudan, L., 'Why was the Logic of Discovery Abandoned?' in T. Nickles (ed.), *Scientific Discovery, Logic, and Rationality*, D. Reidel, Dordrecht, 1980.

Laudan, L., 'A Problem-Solving Approach to Scientific Progress', in I. Hacking (ed.), *Scientific Revolutions*, OUP, Oxford, 1981a.

Laudan, L., 'A Confutation of Convergent Realism', *Philosophy of Science* **48** (1981b), 19—49.

Laudan, L., *Science and Values*, University of California Press, Berkeley, Los Angeles, London, 1984.

Leplin, J., 'Reference and Scientific Realism', *Stud. Hist. Phil. Sci.* **10** (1979), 265—284.

Levin, M. E., 'On Theory-Change and Meaning-Change', *Phil. Sci.* **46** (1979), 407—424.

Ludwig, G., *Die Grundstrukturen einer physikalischen Theorie*, Springer-Verlag, Berlin, 1978.

Makowsky, J. A., Shelah, S. and Stavi, J., 'Δ-Logics and Generalized Quantifiers', *Ann. Math. Log.* **10** (1976), 155—192.

Maksimova, L. L., 'Interpolation Properties of Superintuitionistic, Positive and Modal Logics', in I. Niiniluoto and E. Saarinen (eds), *Intensional Logic: Theory and Applications* (*Acta Philosophica Fennica* **35**), Helsinki, 1982.

Manders, M., 'On the Space—Time Ontology of Physical Theories', *Phil. Sci.* **49** (1982), 575—590.

Malhas, O. Q., 'Space—Time Geometries for One-Dimensional Space', in W. Balzer *et al.* (eds), 1984.

Mayr, D., 'Investigations of the Concept of Reduction, I', *Erkenntnis* **10** (1976), 275—294.

Mayr, D., 'Investigations of the Concept of Reduction, II', *Erkenntnis* **16** (1981), 109—129.

Moulines, C.-U., 'Intertheoretic Approximation: the Kepler—Newton Case', *Synthese* **45** (1980), 387—412.

Mundy, B., 'Relational Theories of Euclidean Space and Minkowski Space-time', *Phil. Sci.* **50** (1983), 205—226.

Nagel, E., 'The Logic of Reduction in the Sciences', *Erkenntnis* **5** (1935), 46—51.

Nagel, E., 'Reduction and Autonomy in the Sciences', *Actes du Huitième Congrès Internationale de Philosophie*, Prague, 1936, pp. 181—186.

Nagel, E., 'The Meaning of Reduction in the Natural Sciences', in R. C. Stauffer (ed.), *Science and Civilization*, University of Wisconsin Press, Madison, 1949.

Nagel, E., *The Structure of Science*, RKP, London, 1961.

Nagel, E., 'Issues in the Logic of Reductive Explanations', in H. Kiefer and M. Munits (eds), *Mind, Science, and History*, SUNY Press, Albany, 1970.

Narens, L., 'Measurement without Archimedean Axioms', *Phil. Sci.* **41** (1974), 374—393.

Niiniluoto, I., *Is Science Progressive?* D. Reidel, Dordrecht, 1984.

Niiniluoto, I. and Tuomela, R. (eds), *The Logic and Epistemology of Scientific Change* (*Acta Philosophica Fennica* **30**), Amsterdam, 1979.

Oikkonen, J., 'The Δ-Extension of $L_{\kappa\kappa}$', Reports from the Dept of Mathematics, University of Helsinki, 1985.

Papineau, D., *Theory and Meaning*, Clarendon Press, Oxford, 1979.

Pearce, D., *Translation, Reduction and Equivalence: Some Topics in Intertheory Relations*, University of Sussex, 1979; Verlag Peter Lang, Bern/Frankfurt, 1985.

Pearce, D., 'Is there Any Theoretical Justification for a Nonstatement View of Theories?', *Synthese* **46** (1981), 1—39.

Pearce, D., 'Stegmueller on Kuhn and Incommensurability', *Brit. J. Phil. Sci.* **33** (1982a), 389—396.

Pearce, D., 'Logical Properties of the Structuralist Concept of Reduction', *Erkenntnis* **18** (1982b), 307—333.

Pearce, D., 'Logical Criteria of Reducibility: a Reply to W. Hoering', in I. Patoluoto, E. Saarinen and P. Stenman (eds), *Vexing Questions* (Reports from the Dept of Philosophy, University of Helsinki, No. 3), 1983.

Pearce, D., 'Research Traditions, Incommensurability and Scientific Progress', *Z. allgem. Wissenschafts.* **15** (1984), 261—271.

Pearce, D., 'Remarks on Physicalism and Reductionism', in G. Holmström and A. Jones (eds), *Action, Logic and Social Theory* (*Acta Philosophica Fennica* **38**), Helsinki, 1985.

Pearce, D., 'Incommensurability and Reduction Reconsidered', *Erkenntnis* **24** (1986), 293—308.

Pearce, D. and Rantala, V., 'Realism and Formal Semantics', *Synthese* **52** (1982a), 39—53.

Pearce, D. and Rantala, V., 'Realism and Reference: some Comments on Putnam', *Synthese* **52** (1982b), 439—448.

Pearce, D. and Rantala, V., 'New Foundations for Metascience', *Synthese* **56** (1983a), 1—26.

Pearce, D. and Rantala, V., 'The Logical Study of Symmetries in Scientific Change', in P. Weingartner and H. Czermak (eds), *Epistemology and Philosophy of Science*, Hölder-Pichler-Tempsky, Vienna, 1983b.

Pearce, D. and Rantala, V., 'Logical Aspects of Scientific Reduction', in P. Weingartner and H. Czermak (eds), 1983c.

Pearce, D. and Rantala, V., 'Constructing General Models of Theory Dynamics', *Studia Logica* **42** (1983d), 347—362.

Pearce, D. and Rantala, V., 'Correspondence as an Intertheory Relation', *Studia Logica* **42** (1983e), 363—371.

Pearce, D. and Rantala, V., 'A Logical Study of the Correspondence Relation', *J. Phil. Logic* **13** (1984a), 47—84.

Pearce, D. and Rantala, V., 'Limiting Case Correspondence between Physical Theories', in Balzer *et al.*, 1984b.

Pearce, D. and Rantala, V., 'Scientific Change, Continuity and Problem Solving', *Philosophia Naturalis* **21** (1984c), 389—399.

Pearce, D. and Rantala, V., 'Approximative Explanation is Deductive-Nomological', *Phil. Sci.* **52** (1985), 126—140.

Pearce, D. and Rantala, V., *Correspondence. A Study in the Semantics of Scientific Change* (to appear), 1988.

Pearce, D. and Tucci, M., 'A General Net Structure for Theoretical Economics', in W. Stegmüller, W. Balzer and W. Spohn (eds), *Philosophy of Economics*, Springer-Verlag, Berlin, Heidelberg, New York, 1982.

Pillay, A., *Gaifman Operations, Minimal Models and the Number of Countable Models*, Dissertation, Bedford College, London, 1977.

Pillay, A., '$\aleph_0$-Categoricity over a Predicate', Manuscript, 1982.

Prawitz, D., 'Meaning and Proofs: on the Conflict between Classical and Intuitionistic Logic', *Theoria* **43** (1977), 2—40.

Przełęcki, M., *The Logic of Empirical Theories*, RKP, London, 1969.

Przełęcki, M., 'Fuzziness as Multiplicity', *Erkenntnis* **10** (1976), 371—380.

Przełęcki, M., 'Commensurable Referents of Incommensurable Terms', in I. Niiniluoto and R. Tuomela (eds), 1979.

Przełęcki, M., 'Conceptual Continuity through Theory Changes', in R. Hilpinen (ed.), *Rationality in Science*, D. Reidel, Dordrecht, 1980.

Putnam, H., 'How Not to Talk About Meaning', in R. S. Cohen and M. Wartofsky (eds), *In Honour of Philipp Frank*, Humanities Press, New York, 1965.

Putnam, H., 'Explanation and Reference', in G. Pearce and P. Maynard (eds), *Conceptual Change*, D. Reidel, Dordrecht, 1973; also in H. Putnam, 1975b.

Putnam, H., 'The Meaning of "Meaning"', in K. Gunderson (ed.), *Language, Mind and Knowledge*, University of Minnesota Press, Minneapolis, 1975a; also in Putnam, 1975b.

Putnam, H., *Philosophical Papers*, Vol. 2, CUP, Cambridge, 1975b.

Putnam, H., 'Realism and Reason', reprinted in H. Putnam, *Meaning and the Moral Sciences*, RKP, London, 1978.

Putnam, H., 'Models and Reality', *J. Symbol. Logic* **45** (1980), 464—482.

Putnam, H., *Reason, Truth and History*, Cambridge University Press, Cambridge, 1981.

Rantala, V., 'The Old and the New Logic of Metascience', *Synthese* **39** (1978), 233—247.

Rantala, V., 'Correspondence and Non-standard Models: a Case Study', in I. Niiniluoto and R. Tuomela (eds), 1979.

Rantala, V., 'On, the Logical Basis of the Structuralist Philosophy of Science', *Erkenntnis* **15** (1980), 269—286.

Schaffner, K., 'Approaches to Reduction', *Phil. Sci.* **34** (1967), 137—147.

Scheffler, I., *Science and Subjectivity*, Bobbs-Merril, Indianapolis, 1967.

Scheibe, E., 'Zum Theorienvergleich in der Physik', in K. M. Meyer-Abich (ed.), *Physik, Philosophie und Politik. Festschrift für C. F. von Weizsäcker*, Carl Hanser Verlag, Munich, 1982.

Scheibe, E., 'Two Types of Successor Relations between Theories', *Z. allgem. Wissenschafts.* **14** (1983), 68—80.

Schmidt, H.-J., 'Tangent Embedding — a Special Kind of Approximate Reduction', in W. Balzer *et al.*, 1984.

Searle, J., *Minds, Brains and Science*, Harvard University Press, Cambridge, 1984.

Sette, A. and Szczerba, L. W., 'Algebraic Characterization of Interpretability', *Notas e Comunicacoes de Matematica* No. 94, Universidade Federal de Pernambuco, 1978.

Sette, A. and Szczerba, L. W., 'Characterisation of Elementary Interpretations in Category Theory', Manuscript, 1983.

Siegel, H., 'Objectivity, Rationality, Incommensurability, and More', *Brit. J. Phil. Sci.* **31** (1980), 359—375.

Sintonen, M., *The Pragmatics of Scientific Explanation* (*Acta Philosophica Fennica* **37**), Helsinki, 1984.

Smorynski, C. A., 'Applications of Kripke Models', in A. Troelstra (ed.), *Metamathematical Investigations of Intuitionistic Arithmetic and Analysis, Lecture Notes in Mathematics*, **344**, Springer-Verlag, Berlin, 1975.

Sneed, J. D., *The Logical Structure of Mathematical Physics*, D. Reidel, Dordrecht, 1971.

Stegmüller, W., *Theorienstrukturen und Theoriendynamik*, Springer-Verlag, New York, Heidelberg, Berlin, 1973 (English translation: *The Structure and Dynamics of Theories*, Springer-Verlag, 1976).

Stegmüller, W., 'Structures and Dynamics of Theories: some Reflections on J. D. Sneed and T. S. Kuhn', *Erkenntnis* **9** (1975), 75—100.

Stegmüller, W., *The Structuralist View of Theories*, Springer-Verlag, Berlin, Heidelberg, New York, 1979.

Stegmüller, W., *Theorie und Erfahrung, Dritter Teilband*, Springer-Verlag, Berlin, etc., 1986.

Suppes, P., *Introduction to Logic*, Van Nostrand, New York, 1957.

Szczerba, L. W., 'Interpretability of Elementary Theories', in R. Butts and J. Hintikka (eds.), *Logic, Foundations of Mathematics and Computability Theory*, D. Reidel, Dordrecht, 1977.

Tennant, N., 'Beth's Theorem and Reductionism', *Pacific Philosophical Quarterly* **66** (1985a), 342—354.

Tennant, N., 'Minimal Logic is Adequate for Popperian Science', *Brit. J. Phil. Sci.* **36** (1985b), 325—329.

Tennant, N., *Anti-realism and Logic, Vol. I: Truth as Eternal*, OUP, Oxford, 1987.

Williams, P. M., 'On the Logical Relations between Expressions of Different Theories', *Brit. J. Phil. Sci.* **24** (1973), 357—367.

Yoshida, R. M., *Reduction in the Physical Sciences*, Dalhousie University Press, Halifax, 1977.

Zahar, E., 'Logic of Discovery or Psychology of Invention?', *Brit. J. Phil. Sci.* **34** (1983), 243—261.

NAME INDEX

247

SUBJECT INDEX

SYNTHESE LIBRARY

Studies in Epistemology, Logic, Methodology,
and Philosophy of Science

Managing Editor:
JAAKKO HINTIKKA, Florida State University, Tallahassee

Editors:
DONALD DAVIDSON, University of California, Berkeley
GABRIËL NUCHELMANS, University of Leyden
WESLEY C. SALMON, University of Pittsburgh

1. J. M. Bochénski, *A Precis of Mathematical Logic.* 1959.
2. P. L. Guiraud, *Problèmes et méthodes de la statistique linguistique.* 1960.
3. Hans Freudenthal (ed.), *The Concept and the Role of the Model in Mathematics and Natural and Social Sciences.* 1961.
4. Evert W. Beth, *Formal Methods. An Introduction to Symbolic Logic and the Study of Effective Operations in Arithmetic and Logic.* 1962.
5. B. H. Kazemier and D. Vuysje (eds.), *Logic and Language. Studies Dedicated to Professor Rudolf Carnap on the Occasion of His Seventieth Birthday.* 1962.
6. Marx W. Wartofsky (ed.), *Proceedings of the Boston Colloquium for the Philosophy of Science 1961–1962.* Boston Studies in the Philosophy of Science, Volume I. 1963.
7. A. A. Zinov'ev, *Philosophical Problems of Many-Valued Logic.* 1963.
8. Georges Gurvitch, *The Spectrum of Social Time.* 1964.
9. Paul Lorenzen, *Formal Logic.* 1965.
10. Robert S. Cohen and Marx W. Wartofsky (eds.), *In Honor of Philipp Frank.* Boston Studies in the Philosophy of Science, Volume II. 1965.
11. Evert W. Beth, *Mathematical Thought. An Introduction to the Philosophy of Mathematics.* 1965.
12. Evert W. Beth and Jean Piaget, *Mathematical Epistemology and Psychology.* 1966.
13. Guido Küng, *Ontology and the Logistic Analysis of Language. An Enquiry into the Contemporary Views on Universals.* 1967.
14. Robert S. Cohen and Marx W. Wartofsky (eds.), *Proceedings of the Boston Colloquium for the Philosophy of Sciences 1964–1966. In Memory of Norwood Russell Hanson.* Boston Studies in the Philosophy of Science, Volume III. 1967.
15. C. D. Broad, *Induction, Probability, and Causation. Selected Papers.* 1968.
16. Günther Patzig, *Aristotle's Theory of the Syllogism. A Logical-Philosophical Study of Book A of the Prior Analytics.* 1968.
17. Nicholas Rescher, *Topics in Philosophical Logic.* 1968.
18. Robert S. Cohen and Marx W. Wartofsky (eds.), *Proceedings of the Boston Colloquium for the Philosophy of Science 1966–1968.* Boston Studies in the Philosophy of Science, Volume IV. 1969

19. Robert S. Cohen and Marx W. Wartofsky (eds.), *Proceedings of the Boston Colloquium for the Philosophy of Science 1966–1968.* Boston Studies in the Philosophy of Science, Volume V. 1969

20. J. W. Davis, D. J. Hockney, and W. K. Wilson (eds.), *Philosophical Logic.* 1969

21. D. Davidson and J. Hintikka (eds.), *Words and Objections. Essays on the Work of W. V. Quine.* 1969.

22. Patrick Suppes. *Studies in the Methodology and Foundations of Science. Selected Papers from 1911 to 1969.* 1969

23. Jaakko Hintikka, *Models for Modalities. Selected Essays.* 1969

24. Nicholas Rescher *et al.* (eds.), *Essays in Honor of Carl G. Hempel. A Tribute on the Occasion of His Sixty-Fifth Birthday.* 1969

25. P. V. Tavanec (ed.), *Problems of the Logic of Scientific Knowledge.* 1969

26. Marshall Swain (ed.), *Induction, Acceptance, and Rational Belief.* 1970.

27. Robert S. Cohen and Raymond J. Seeger (eds.), *Ernst Mach: Physicist and Philosopher.* Boston Studies in the Philosophy of Science, Volume VI. 1970.

28. Jaakko Hintikka and Patrick Suppes, *Information and Inference.* 1970.

29. Karel Lambert, *Philosophical Problems in Logic. Some Recent Developments.* 1970.

30. Rolf A. Eberle, *Nominalistic Systems.* 1970.

31. Paul Weingartner and Gerhard Zecha (eds.), *Induction, Physics, and Ethics.* 1970.

32. Evert W. Beth, *Aspects of Modern Logic.* 1970.

33. Risto Hilpinen (ed.), *Deontic Logic: Introductory and Systematic Readings.* 1971.

34. Jean-Louis Krivine, *Introduction to Axiomatic Set Theory.* 1971.

35. Joseph D. Sneed, *The Logical Structure of Mathematical Physics.* 1971.

36. Carl R. Kordig, *The Justification of Scientific Change.* 1971.

37. Milic Capek, *Bergson and Modern Physics.* Boston Studies in the Philosophy of Science, Volume VII. 1971.

38. Norwood Russell Hanson, *What I Do Not Believe, and Other Essays* (ed. by Stephen Toulmin and Harry Woolf). 1971.

39. Roger C. Buck and Robert S. Cohen (eds.), *PSA 1970. In Memory of Rudolf Carnap.* Boston Studies in the Philosophy of Science, Volume VIII. 1971

40. Donald Davidson and Gilbert Harman (eds.), *Semantics of Natural Language.* 1972.

41. Yehoshua Bar-Hillel (ed.), *Pragmatics of Natural Languages.* 1971.

42. Sören Stenlund, *Combinators, λ-Terms and Proof Theory.* 1972.

43. Martin Strauss, *Modern Physics and Its Philosophy. Selected Paper in the Logic, History, and Philosophy of Science.* 1972.

44. Mario Bunge, *Method, Model and Matter.* 1973.

45. Mario Bunge, *Philosophy of Physics.* 1973.

46. A. A. Zinov'ev, *Foundations of the Logical Theory of Scientific Knowledge (Complex Logic).* (Revised and enlarged English edition with an appendix by G. A. Smirnov, E. A. Sidorenka, A. M. Fedina, and L. A. Bobrova.) Boston Studies in the Philosophy of Science, Volume IX. 1973.

47. Ladislav Tondl, *Scientific Procedures.* Boston Studies in the Philosophy of Science, Volume X. 1973.

48. Norwood Russell Hanson, *Constellations and Conjectures* (ed. by Willard C. Humphreys, Jr.). 1973

49. K. J. J. Hintikka, J. M. E. Moravcsik, and P. Suppes (eds.), *Approaches to Natural Language.* 1973.

50. Mario Bunge (ed.), *Exact Philosophy – Problems, Tools, and Goals.* 1973.
51. Radu J. Bogdan and Ilkka Niiniluoto (eds.), *Logic, Language, and Probability.* 1973.
52. Glenn Pearce and Patrick Maynard (eds.), *Conceptual Change.* 1973.
53. Ilkka Niiniluoto and Raimo Tuomela, *Theoretical Concepts and Hypothetico-Inductive Inference.* 1973.
54. Roland Fraissé, *Course of Mathematical Logic* – Volume 1: *Relation and Logical Formula.* 1973.
55. Adolf Grünbaum, *Philosophical Problems of Space and Time.* (Second, enlarged edition.) Boston Studies in the Philosophy of Science, Volume XII. 1973.
56. Patrick Suppes (ed.), *Space, Time, and Geometry.* 1973.
57. Hans Kelsen, *Essays in Legal and Moral Philosophy* (selected and introduced by Ota Weinberger). 1973.
58. R. J. Seeger and Robert S. Cohen (eds.), *Philosophical Foundations of Science.* Boston Studies in the Philosophy of Science, Volume XI. 1974.
59. Robert S. Cohen and Marx W. Wartofsky (eds.), *Logical and Epistemological Studies in Contemporary Physics.* Boston Studies in the Philosophy of Science, Volume XIII. 1973.
60. Robert S. Cohen and Marx W. Wartofsky (eds.), *Methodological and Historical Essays in the Natural and Social Sciences. Proceedings of the Boston Colloquium for the Philosophy of Science 1969–1972.* Boston Studies in the Philosophy of Science, Volume XIV. 1974.
61. Robert S. Cohen, J. J. Stachel, and Marx W. Wartofsky (eds.), *For Dirk Struik. Scientific, Historical and Political Essays in Honor of Dirk J. Struik.* Boston Studies in the Philosophy of Science, Volume XV. 1974.
62. Kazimierz Ajdukiewicz, *Pragmatic Logic* (transl. from the Polish by Olgierd Wojtasiewicz). 1974.
63. Sören Stenlund (ed.), *Logical Theory and Semantic Analysis. Essays Dedicated to Stig Kanger on His Fiftieth Birthday.* 1974.
64. Kenneth F. Schaffner and Robert S. Cohen (eds.), *Proceedings of the 1972 Biennial Meeting, Philosophy of Science Association.* Boston Studies in the Philosophy of Science, Volume XX. 1974.
65. Henry E. Kyburg, Jr., *The Logical Foundations of Statistical Inference.* 1974.
66. Marjorie Grene, *The Understanding of Nature. Essays in the Philosophy of Biology.* Boston Studies in the Philosophy of Science, Volume XXIII. 1974.
67. Jan M. Broekman, *Structuralism: Moscow, Prague, Paris.* 1974.
68. Norman Geschwind, *Selected Papers on Language and the Brain,* Boston Studies in the Philosophy of Science, Volume XVI. 1974.
69. Roland Fraissé, *Course of Mathematical Logic* – Volume 2: *Model Theory.* 1974.
70. Andrzej Grzegorczyk, *An Outline of Mathematical. Logic. Fundamental Results and Notions Explained with All Details.* 1974.
71. Franz von Kutschera, *Philosophy of Language.* 1975.
72. Juha Manninen and Raimo Tuomela (eds.), *Essays on Explanation and Understanding. Studies in the Foundations of Humanities and Social Sciences.* 1976.
73. Jaakko Hintikka (ed.), *Rudolf Carnap, Logical Empiricist. Materials and Perspectives.* 1975.
74. Milic Capek (ed.), *The Concepts of Space and Time. Their Structure and Their Development.* Boston Studies in the Philosophy of Science, Volume XXII. 1976.

75. Jaakko Hintikka and Unto Remes, *The Method of Analysis. Its Geometrical Origin and Its General Significance.* Boston Studies in the Philosophy of Science, Volume XXV. 1974.
76. John Emery Murdoch and Edith Dudley Sylla, *The Cultural Context of Medieval Learning.* Boston Studies in the Philosophy of Science, Volume XXVI. 1975.
77. Stefan Amsterdamski, *Between Experience and Metaphysics. Philosophical Problems of the Evolution of Science.* Boston Studies in the Philosophy of Science, Volume XXXV. 1975.
78. Patrick Suppes (ed.), *Logic and Probability in Quantum Mechanics.* 1976.
79. Hermann von Helmholtz: *Epistemological Writings. The Paul Hertz/Moritz Schlick Centenary Edition of 1921 with Notes and Commentary by the Editors.* (Newly translated by Malcolm F. Lowe. Edited, with an Introduction and Bibliography, by Robert S. Cohen and Yehuda Elkana.) Boston Studies in the Philosophy of Science, Volume XXXVII. 1975.
80. Joseph Agassi, *Science in Flux.* Boston Studies in the Philosophy of Science, Volume XXVIII. 1975.
81. Sandra G. Harding (ed.), *Can Theories Be Refuted? Essays on the Duhem-Quine Thesis.* 1976.
82. Stefan Nowak, *Methodology of Sociological Research. General Problems.* 1977.
83. Jean Piaget, Jean-Blaise Grize, Alina Szeminska, and Vinh Bang, *Epistemology and Psychology of Functions.* 1977.
84. Marjorie Grene and Everett Mendelsohn (eds.), *Topics in the Philosophy of Biology.* Boston Studies in the Philosophy of Science, Volume XXVII. 1976.
85. E. Fischbein, *The Intuitive Sources of Probabilistic Thinking in Children.* 1975.
86. Ernest W. Adams, *The Logic of Conditionals. An Application of Probability to Deductive Logic.* 1975.
87. Marian Przelecki and Ryszard Wójcicki (eds.), *Twenty-Five Years of Logical Methodology in Poland.* 1976.
88. J. Topolski, *The Methodology of History.* 1976.
89. A. Kasher (ed.), *Language in Focus: Foundations, Methods and Systems. Essays Dedicated to Yehoshua Bar-Hillel.* Boston Studies in the Philosophy of Science, Volume XLIII. 1976.
90. Jaakko Hintikka, *The Intentions of Intentionality and Other New Models for Modalities.* 1975.
91. Wolfgang Stegmüller, *Collected Papers on Epistemology, Philosophy of Science and History of Philosophy.* 2 Volumes. 1977.
92. Dov M. Gabbay, *Investigations in Modal and Tense Logics with Applications to Problems in Philosophy and Linguistics.* 1976.
93. Radu J. Bodgan, *Local Induction.* 1976.
94. Stefan Nowak, *Understanding and Prediction. Essays in the Methodology of Social and Behavioral Theories.* 1976.
95. Peter Mittelstaedt, *Philosophical Problems of Modern Physics.* Boston Studies in the Philosophy of Science, Volume XVIII. 1976.
96. Gerald Holton and William Blanpied (eds.), *Science and Its Public; The Changing Relationship.* Boston Studies in the Philosophy of Science, Volume XXXIII. 1976.
97. Myles Brand and Douglas Walton (eds.), *Action Theory.* 1976.
98. Paul Gochet, *Outline of a Nominalist Theory of Proposition. An Essay in the Theory of Meaning.* 1980.
99. R. S. Cohen, P. K. Feyerabend, and M. W. Wartofsky (eds.), *Essays in Memory of Imre Lakatos.* Boston Studies in the Philosophy of Science, Volume XXXIX. 1976.

100. R. S. Cohen and J. J. Stachel (eds.), *Selected Papers of Léon Rosenfield*. Boston Studies in the Philosophy of Science, Volume XXI. 1978.

101. R. S. Cohen, C. A. Hooker, A. C. Michalos, and J. W. van Evra (eds.), *PSA 1974: Proceedings of the 1974 Biennial Meeting of the Philosophy of Science Association*. Boston Studies in the Philosophy of Science, Volume XXXII. 1976.

102. Yehuda Fried and Joseph Agassi, *Paranoia: A Study in Diagnosis*. Boston Studies in the Philosophy of Science, Volume L. 1976.

103. Marian Przelecki, Klemens Szaniawski, and Ryszard Wójcicki (eds.), *Formal Methods in the Methodology of Empirical Sciences*. 1976.

104. John M. Vickers, *Belief and Probability*. 1976.

105. Kurt H. Wolff, *Surrender and Catch: Experience and Inquiry Today*. Boston Studies in the Philosophy of Science, Volume LI. 1976.

106. Karel Kosík, *Dialectics of the Concrete*. Boston Studies in the Philosophy of Science, Volume LII. 1976.

107. Nelson Goodman, *The Structure of Appearance* (Third edition.) Boston Studies in the Philosophy of Science, Volume LIII. 1977.

108. Jerzy Giedymin (ed.), *Kazimierz Ajdukiewicz: The Scientific World-Perspective and Other Essays, 1931-1963*. 1978.

109. Robert L. Causey, *Unity of Science*. 1977.

110. Richard E. Grandy, *Advanced Logic for Applications*. 1977.

111. Robert P. McArthur, *Tense Logic*. 1976.

112. Lars Lindahl, *Position and Change. A Study in Law and Logic*. 1977.

113. Raimo Tuomela, *Dispositions*. 1978.

114. Herbert A. Simon, *Models of Discovery and Other Topics in the Methods of Science*. Boston Studies in the Philosophy of Science, Volume LIV. 1977.

115. Roger D. Rosenkrantz, *Inference, Method and Decision*. 1977.

116. Raimo Tuomela, *Human Action and Its Explanation. A Study on the Philosophical Foundations of Psychology*. 1977.

117. Morris Lazerowitz, *The Language of Philosophy. Freud and Wittgenstein*. Boston Studies in the Philosophy of Science, Volume LV. 1977.

119. Jerzy Pelc, *Semiotics in Poland, 1894-1969*. 1978.

120. Ingmar Pörn, *Action Theory and Social Science. Some Formal Models*. 1977.

121. Joseph Margolis, *Persons and Mind. The Prospects of Nonreductive Materialism*. Boston Studies in the Philosophy of Science, Volume LVII. 1977.

122. Jaakko Hintikka, Ilkka Niiniluoto, and Esa Saarinen (eds.), *Essays on Mathematical and Philosophical Logic*. 1978.

123. Theo A. F. Kuipers, *Studies in Inductive Probability and Rational Expectation*. 1978.

124. Esa Saarinen, Risto Hilpinen, Ilkka Niiniluoto, and Merrill Provence Hintikka (eds.), *Essays in Honour of Jaakko Hintikka on the Occasion of His Fiftieth Birthday*. 1978.

125. Gerard Radnitzky and Gunnar Andersson (eds.), *Progress and Rationality in Science*. Boston Studies in the Philosophy of Science, Volume LVIII. 1978.

126. Peter Mittelstaedt, *Quantum Logic*. 1978.

127. Kenneth A. Bowen, *Model Theory for Modal Logic. Kripke Models for Modal Predicate Calculi*. 1978.

128. Howard Alexander Bursen, *Dismantling the Memory Machine. A Philosophical Investigation of Machine Theories of Memory*. 1978.

129. Marx W. Wartofsky, *Models: Representation and the Scientific Understanding*. Boston Studies in the Philosophy of Science, Volume XLVIII. 1979.

130. Don Ihde, *Technics and Praxis. A Philosophy of Technology.* Boston Studies in the Philosophy of Science, Volume XXIV. 1978.
131. Jerzy J. Wiatr (ed.), *Polish Essays in the Methodology of the Social Sciences.* Boston Studies in the Philosophy of Science, Volume XXIX. 1979.
132. Wesley C. Salmon (ed.), *Hans Reichenbach: Logical Empiricist.* 1979.
133. Peter Bieri, Rolf-P. Horstmann, and Lorenz Krüger (eds.), *Transcendental Arguments in Science Essays in Epistemology.* 1979.
134. Mihailo Marković and Gajo Petrović (eds.), *Praxis, Yugoslav Essays in the Philosophy and Methodology of the Social Sciences.* Boston Studies in the Philosophy of Science, Volume XXXVI. 1979.
135. Ryszard Wójcicki, *Topics in the Formal Methodology of Empirical Sciences.* 1979.
136. Gerard Radnitzky and Gunnar Andersson (eds.), *The Structure and Development of Science.* Boston Studies in the Philosophy of Science, Volume LIX. 1979.
137. Judson Chambers Webb, *Mechanism, Mentalism, and Metamathematics. An Essay on Finitism.* 1980.
138. D. F. Gustafson and B. L. Tapscott (eds.), *Body, Mind, and Method. Essays in Honor of Virgil C. Aldrich.* 1979.
139. Leszek Nowak, *The Structure of Idealization. Towards a Systematic Interpretation of the Marxian Idea of Science.* 1979.
140. Chaim Perelman, *The New Rhetoric and the Humanities. Essays on Rhetoric and Its Applications.* 1979.
141. Wlodzimierz Rabinowicz, *Universalizability. A Study in Morals and Metaphysics.* 1979.
142. Chaim Perelman, *Justice, Law, and Argument. Essays on Moral and Legal Reasoning.* 1980.
143. Stig Kanger and Sven Öhman (eds.), *Philosophy and Grammar. Papers on the Occasion of the Quincentennial of Uppsala University.* 1980.
144. Tadeusz Pawlowski, *Concept Formation in the Humanities and the Social Sciences.* 1980.
145. Jaakko Hintikka, David Gruender, and Evandro Agazzi (eds.), *Theory Change, Ancient Axiomatics, and Galileo's Methodology. Proceedings of the 1978 Pisa Conference on the History and Philosophy of Science,* Volume I. 1981.
146. Jaakko Hintikka, David Gruender, and Evandro Agazzi, *Probabilistic Thinking, Thermodynamics, and the Interaction of the History and Philosophy of Science. Proceedings of the 1978 Pisa Conference on the History and Philosophy of Science,* Volume II. 1981.
147. Uwe Mönnich (ed.), *Aspects of Philosophical Logic. Some Logical Forays into Central Notions of Linguistics and Philosophy.* 1981.
148. Dov M. Gabbay, *Semantical Investigations in Heyting's Intuitionistic Logic.* 1981.
149. Evandro Agazzi (ed.), *Modern Logic – A Survey. Historical, Philosophical, and Mathematical Aspects of Modern Logic and Its Applications.* 1981.
150. A. F. Parker-Rhodes, *The Theory of Indistinguishables. A Search for Explanatory Principles below the Level of Physics.* 1981.
151. J. C. Pitt, *Pictures, Images, and Conceptual Change. An Analysis of Wilfrid Sellars' Philosophy of Science.* 1981.
152. R. Hilpinen (ed.), *New Studies in Deontic Logic. Norms, Actions, and the Foundations of Ethics.* 1981.
153. C. Dilworth, *Scientific Progress. A Study Concerning the Nature of the Relation Between Successive Scientific Theories.* 1981.
154. D. W. Smith and R. McIntyre, *Husserl and Intentionality. A Study of Mind, Meaning, and Language.* 1982.

155. R. J. Nelson, *The Logic of Mind*. 1982.
156. J. F. A. K. van Benthem, *The Logic of Time. A Model-Theoretic Investigation into the Varieties of Temporal Ontology, and Temporal Discourse*. 1982.
157. R. Swinburne (ed.), *Space, Time and Causality*. 1982.
158. R. D. Rozenkrantz, *E. T. Jaynes: Papers on Probability, Statistics and Statistical Physics*. 1983.
159. T. Chapman, *Time: A Philosophical Analysis*. 1982.
160. E. N. Zalta, *Abstract Objects. An Introduction to Axiomatic Metaphysics*. 1983.
161. S. Harding and M. B. Hintikka (eds.), *Discovering Reality. Feminist Perspectives on Epistemology, Metaphysics, Methodology, and Philosophy of Science*. 1983.
162. M. A. Stewart (ed.), *Law, Morality and Rights*. 1983.
163. D. Mayr and G. Süssmann (eds.), *Space, Time, and Mechanics. Basic Structure of a Physical Theory*. 1983.
164. D. Gabbay and F. Guenthner (eds.), *Handbook of Philosophical Logic*. Vol. I. 1983.
165. D. Gabbay and F. Guenthner (eds.), *Handbook of Philosophical Logic*. Vol. II. 1984.
166. D. Gabbay and F. Guenthner (eds.), *Handbook of Philosophical Logic*. Vol. III. 1985.
167. D. Gabbay and F. Guenthner (eds.), *Handbook of Philosophical Logic*. Vol. IV, forthcoming.
168. Andrew, J. I. Jones, *Communication and Meaning*. 1983.
169. Melvin Fitting, *Proof Methods for Modal and Intuitionistic Logics*. 1983.
170. Joseph Margolis, *Culture and Cultural Entities*. 1984.
171. Raimo Tuomela, *A Theory of Social Action*. 1984.
172. Jorge J. E. Gracia, Eduardo Rabossi, Enrique Villanueva, and Marcelo Dascal (eds.), *Philosophical Analysis in Latin America*. 1984.
173. Paul Ziff, *Epistemic Analysis. A Coherence Theory of Knowledge*. 1984.
174. Paul Ziff, *Antiaesthetics. An Appreciation of the Cow with the Subtile Nose*. 1984.
175. Wolfgang Balzer, David A. Pearce, and Heinz-Jürgen Schmidt (eds.), *Reduction in Science. Structure, Examples, Philosophical Problems*. 1984.
176. Aleksander Peczenik, Lars Lindahl, and Bert van Roermund (eds.), *Theory of Legal Science. Proceedings of the Conference on Legal Theory and Philosophy of Science, Lund, Sweden, December 11–14, 1983*. 1984.
177. Ilkka Niiniluoto, *Is Science Progressive?* 1984.
178. Binal Matilal and Jaysankar Lal Shaw (eds.), *Exploratory Essays in Current Theories and Classical Indian Theories of Meaning and Reference*. 1985.
179. Peter Kroes, *Time: Its Structure and Role in Physical Theories*. 1985.
180. James H. Fetzer, *Sociobiology and Epistemology*, 1985.
181. L. Haaparanta and J. Hintikka, *Frege Synthesized. Essays on the Philosophical and Foundational Work of Gottlob Frege*. 1986.
182. Michael Detlefsen, *Hilbert's Program. An Essay on Mathematical Instrumentalism*. 1986.
183. James L. Golden and Joseph J. Pilotta (eds.), *Practical Reasoning in Human Affairs. Studies in Honor of Chaim Perelman*. 1986.
184. Henk Zandvoort, *Models of Scientific Development and the Case of Nuclear Magnetic Resonance*. 1986.
185. Ilkka Niiniluoto, *Truthlikeness*. 1987
186. Wolfgang Balzer, C. Ulises Moulines, and Joseph D. Sneed, *An Architectonic for Science*. 1987